WOMEN AND GIS VOL. 3

Champions of a Sustainable World

Written by **Esri Press**

Esri Press
REDLANDS | CALIFORNIA

Esri Press, 380 New York Street, Redlands, California 92373-8100
Copyright © 2021 Esri
All rights reserved.
Printed in the United States of America
25 24 23 22 21   1 2 3 4 5 6 7 8 9 10

ISBN: 9781589486379
Library of Congress Control Number: 2021934428

For purchasing and distribution options (both domestic and international), please visit esripress.esri.com.

DEDICATION

As in the first two volumes of *Women and GIS*, this book is dedicated to the women working here at Esri®. They continue to inspire us and the people around them with their knowledge, hard work, and dedication to making the world a better place through science and GIS. It's a pleasure to work with you.

CONTENTS

Foreword xiii

Preface xix

 PRISCILLA MBAMA ABASI   3
Making maps and flying drones to save lives

 ARIANNA ARMELLI   8
Taking a business risk to help others avoid risk

 MARYGRACE BALINOS   16
From imagined cities to real solutions

 FIONA BECKER   22
Blending the beauty of science and art

 MAGGIE CAWLEY   33
Traveling the open road for open data

 HANAN DARWISHE   43
Reaching for the stars with her feet on the ground

 ELENA FIELD   48
Charting the unknown in the Antarctic

GABI FLEURY   54
Forging a path to coexistence with wildlife

AFRICA FLORES-ANDERSON   60
Fighting for a sustainable world, from Guatemala to the Himalayas

MIRIAM GONZÁLEZ   65
Democratizing access to geospatial data

HEALY HAMILTON   71
Answering life's call to help save the diversity of life

KATHARINE HAYHOE   78
Spreading the word on climate change—and action

JACQUE LARRAINZAR   85
Mapping a city's path to racial equity

ANNITA LUCCHESI   90
Carving out space for Indigenous mapping

SAVANNA NAGORSKI & MELISSA K. SCHUTTEN   96

Supporting urban development and tribal communities

TRISALYN NELSON   109
Turning geography into practical solutions

LINDA OCHWADA   114
Leading the way on geospatial AI and innovation in Africa

ZARITH PINEDA **120**
Generating empathy through equitable design

MAYA QUIÑONES **126**
Bringing forestry data to life in the Caribbean

ALICE RATHJEN **133**
Going on a spiritual journey to map genomes

MARIA-ALICIA SERRANO **139**
Bridging communities using insights from GIS

ALINA SHEMETOVA **145**
Energizing GIS from a legacy of science

ARIELLE SIMMONS-STEFFEN **153**
Protecting watersheds for generations to come

LAUREN SINCLAIR **162**
Empowering kids using GIS

REGAN SMYTH **170**
Seeing the big picture and keeping it real

PATRICIA SOLIS **179**
Serving as an ambassador for people, places, and peace

NAVYA TRIPATHI **189**
Pioneering the future of GIS

KALPANA VISWANATH 194
Pinning her business on the safety of cities

JULIA WAGEMANN 199
Expanding the network of female leaders in GIS

FAUSTINE WILLIAMS 205
Improving health outcomes for underserved populations

About the Esri Press team 211

KEY

The individuals profiled in this book work in a variety of fields. Use this key and the sunburst icon beside each name to learn about that person's top three fields.

Secondary field

Primary field

Tertiary field

 Science and Research

 Education and the Arts

Business and Entrepreneurship

Conservation and the Environment

 Social Justice

 Humanitarianism

On the next page, view this dataset sorted by primary field.

Science and Research

Business and Entrepreneurship

Julia Wagemann

Patricia Solis

Trisalyn Nelson

Elena Field

Hanan Darwishe

Maya Quiñones

Katharine Hayhoe

Maria-Alicia Serrano

Zarith Pineda

Navya Tripathi

Faustine Williams

Alice Rathjen

Arianna Armelli

Linda Ochwada

Kalpana Viswanath

Savanna Nagorski

Priscilla Mbama Abasi

Marygrace Balinos

Alina Shemetova

Annita Lucchesi

Mapping their Fields

Social Justice

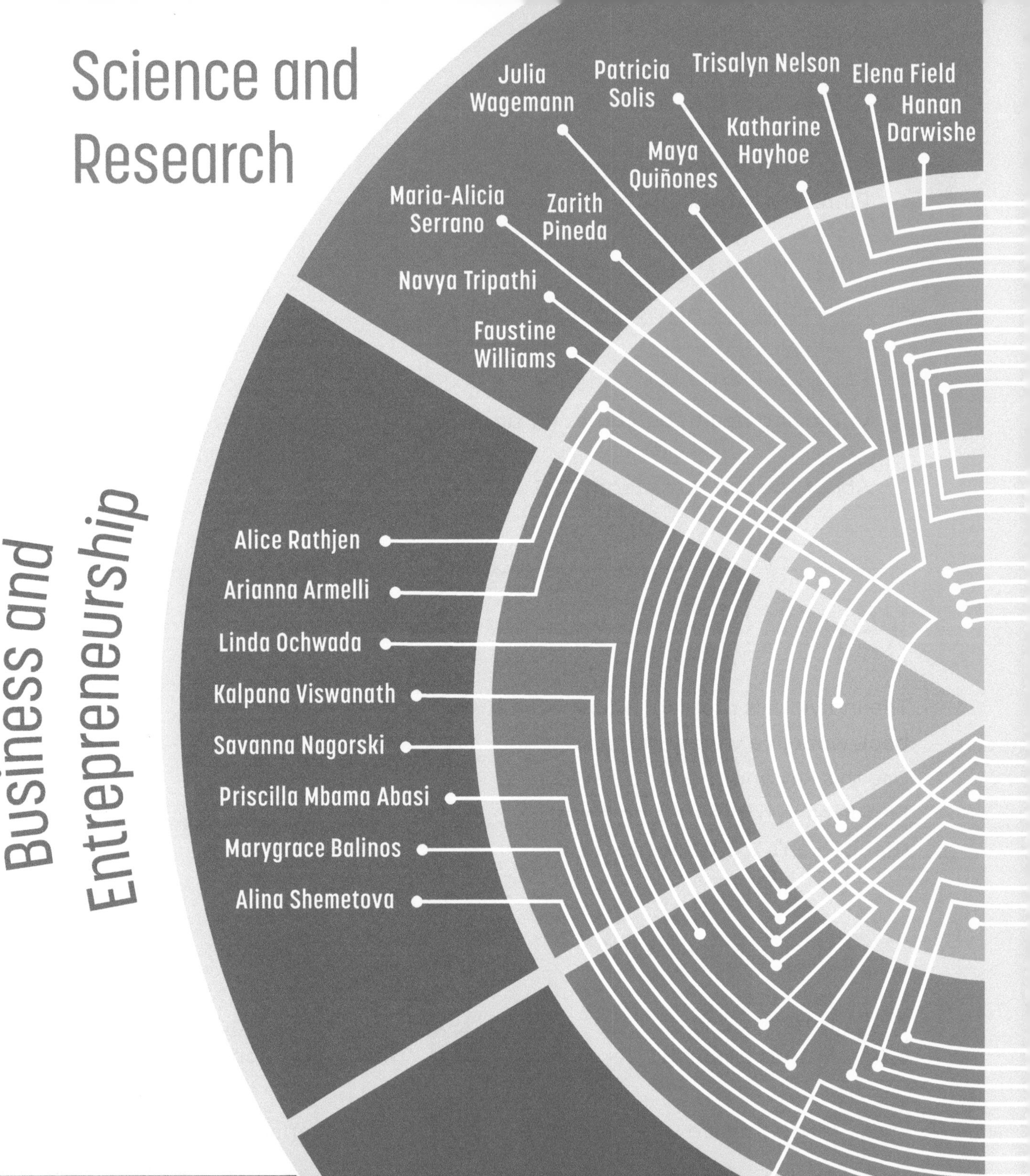

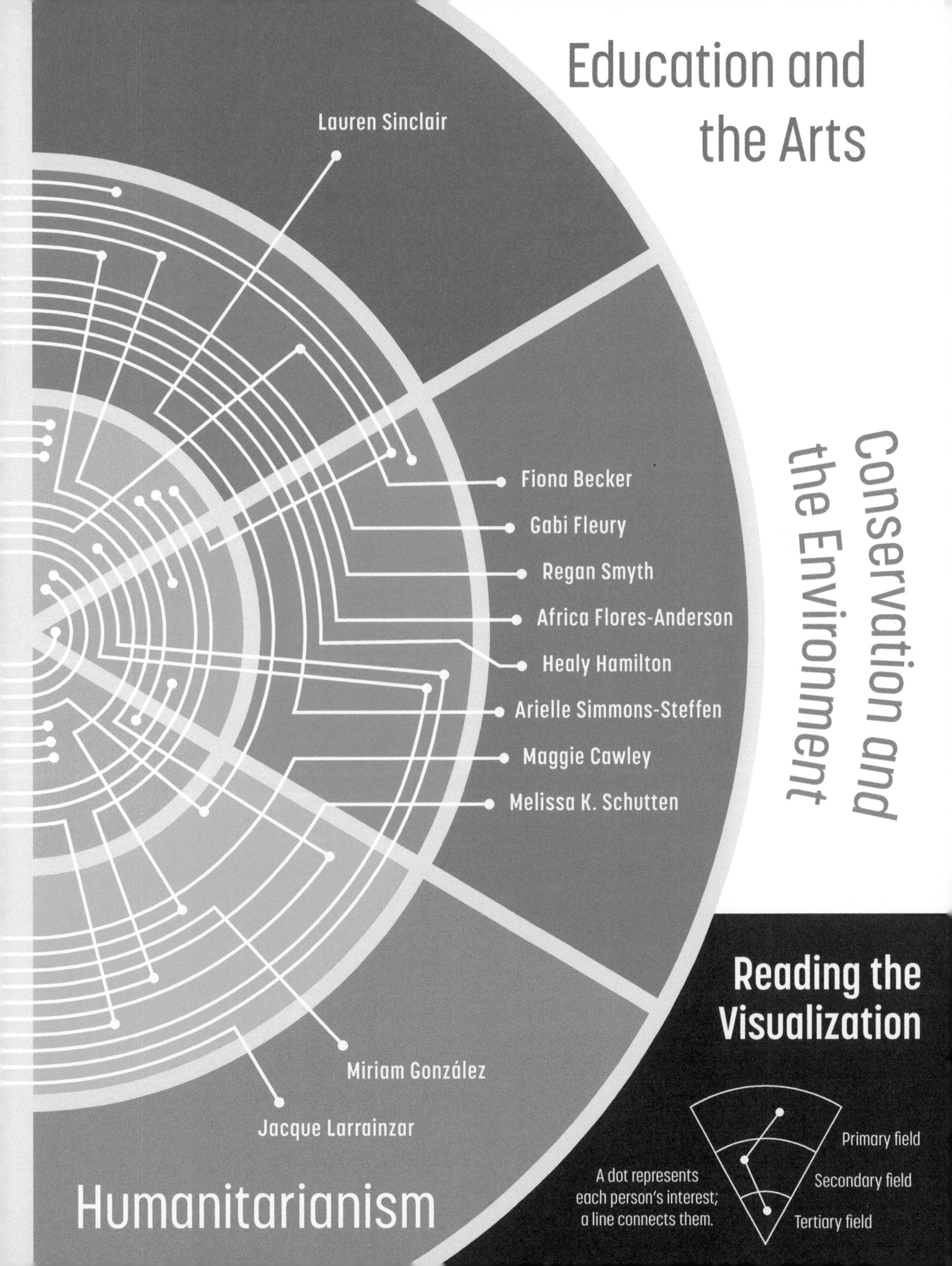

Education and the Arts

Conservation and the Environment

Lauren Sinclair

Fiona Becker

Gabi Fleury

Regan Smyth

Africa Flores-Anderson

Healy Hamilton

Arielle Simmons-Steffen

Maggie Cawley

Melissa K. Schutten

Miriam González

Jacque Larrainzar

Humanitarianism

Reading the Visualization

A dot represents each person's interest; a line connects them.

Primary field

Secondary field

Tertiary field

FOREWORD

From *Women and GIS, Volume 2: Stars of Spatial Science*

I had a passion for the natural world from a very young age. My mother nurtured this passion by finding books for me to read. She thought wisely, "If I get books that Jane is interested in, she'll learn to read more quickly." And of course, she was right.

When I was eight years old, I "met" Doctor Doolittle. I loved the story in which he took animals from the circus back to Africa. Two years later, in the secondhand bookshop where I spent hours every Saturday, I found a little book called *Tarzan of the Apes*. We had very little money, but I used to save up my few pennies of pocket money, and I had just enough to buy that book. And I fell in love with Tarzan. And what did he do? He married the *wrong* Jane. I was jealous! I thought she was a wimp and that I would have made a much better mate! Of course, I knew there wasn't a real Tarzan, but that was when my dream began—I would go to Africa, live with wild animals, and write books about them.

Everyone I told laughed at me. How could I do that? We had very little money. World War II was raging. Africa was far away. And I was "just a girl." And girls didn't do that sort of thing. So they told me I should dream about something I could actually achieve and forget about going to Africa. When a career counselor came to the school and heard that I wanted to go out and study animals in the wild, she laughed too. She suggested instead that I consider becoming a photographer and make portraits of people's pet dogs and cats. There was no suggestion that I become a scientist studying the behaviour of African animals, because no women were pursuing such a path. In fact, very few women were trying to be scientists of any sort.

But throughout this, my mother always offered encouragement. "If you really want to do this, you're going to have to work really hard. Take advantage of every opportunity, don't give up, and you will find a way," she said.

As I look back on my life and think of all the amazing people who have supported me, the person to whom I owe the most, who was the greatest inspiration, who helped me to be what I am today was my wonderful mother. Right from the beginning, she supported my passion for animals. When I was 18 months old, she came into my room one night to find I had taken a handful of earthworms to bed with me. Instead of saying, "Ugh! Take those dirty things out of your bed," she said, "If you leave them here, they'll die. They need the earth." So together we gathered them up and returned them to the garden. Then, when I was four and a half years old, she took me for a holiday to a farm (a *proper* one, not a factory farm). I still remember meeting cows, horses, pigs out in the fields. I was given a job: collect the eggs. The hens pecked around in the farmyard but laid their eggs in nest boxes placed around a number of wooden huts where they slept at night. I asked, "Where is the hole where the egg comes out?" I couldn't see a big enough one! No one told me. I remember crawling after a hen going into one of the houses, hoping I could solve the mystery for myself, I suppose. But with squawks of fear, she flew out. I must have realized that no hen would lay an egg there, so I went into an empty henhouse and waited. I was on the path of discovery. I was gone for four hours. My mother did not know where I was. It was getting dark when she saw an excited little girl rushing towards the house. She must have been scared, but instead of reprimanding me—"How dare you go off without telling me?"—she sat down to hear my story. I tell it now because it shows I had the makings of a scientist: curiosity, asking questions, deciding to find out for myself, making a mistake, and learning patience. It was all there, and a different kind of mother might have crushed that scientific curiosity and I might not have become who I am today.

I did well at school but there was no money for university. I learned to do shorthand and typing because I had to get a job of some sort. I worked in London for a couple of years—and then came a letter inviting me to Kenya to stay with a school friend. Opportunity! I went home and got a job as a waitress in a hotel around the corner. It was hard work, but after about five months, I had saved up enough for a return fare to Kenya. I was 23 years old when I set off by boat. What an adventure! I was fortunate enough to meet Dr. Louis Leakey, the famous paleontologist. He was impressed by how much I knew about African animals—I had read everything I could about them. It was he who asked if I was prepared to study wild chimpanzees. How amazing. I would be living with and learning from not just any animal but the one closest to us! Eventually, Louis found an American philanthropist prepared to support the crazy plan—sending a young woman into the forest with no experience and no degree. So, in 1960, I travelled to Gombe Stream National Park in Tanzania to study chimpanzees. Soon I realized how like us they are. The breakthrough observation was seeing a chimpanzee, whom I had named David Greybeard, using and making grass and twig tools to fish for termites. Previously, scientists believed that only humans made tools—we were defined as "man the toolmaker"—leading Louis Leakey to say, "We shall now have to redefine tool, redefine man, or accept chimpanzees as humans."

In chimpanzee society, some females are much better than others, and we now know, after spending 60 years studying the same communities, that the offspring of mothers who were affectionate, protective but not overprotective, and, above all, supportive tend to be more assertive, have more self-confidence. The males achieve a high rank in the hierarchy and the females make better mothers

themselves. Throughout our evolution, it was important for females to be good mothers; they needed to be patient, quick to understand the wants and needs of their infants before they could speak, and good at keeping the peace between family members. If these qualities are, to some extent, handed down in our female genes, this may explain why women so often make such good observers.

Over the years at Gombe, whilst continuing the observations of individual chimpanzees, we have introduced many of the new technologies that have transformed the way that many field biologists work. When I began, I was equipped with nothing more than binoculars, notebook, and pencil. My first notes were written up by hand. In 1962 I was given a manual typewriter. Next I got a small grant to get a lightweight telescope so that I could observe chimpanzees over a greater distance, and then a small tape recorder so that observations could be made in greater detail and typed out later.

In 1961 the National Geographic Society sent out Hugo van Lawick to film the chimpanzees—with an old Bolex camera. By that time, I had established a small field-research station, and with several students observing and recording, and Hugo's film footage and still photos, we began to amass a large amount of data.

Other developments we were able to employ over the years included infrared technology, lightweight video-recording equipment, and DNA profiling. Each new technology we implement opens up whole new avenues of research and understanding.

One of the greatest technology contributions to our conservation efforts has been from the use of geographic information systems (GIS) through our partnership with Esri. We use the latest GIS technology to determine the range of the chimpanzees and we use satellite imagery to assess the impact of Tacare—our program to help villagers find ways of making a living without destroying their environment. The satellite imagery obtained over the years has allowed us to study how the country around Gombe, once completely deforested, has gradually seen more and more of the bare hills again covered in forest. Not only is GIS technology helping us understand chimpanzee spatial behavior, it is also providing a window to understand what's happening in terms of conservation and is key to the scaling of our community-based conservation efforts beyond the local level.

Let me end by repeating those important words of wisdom from my mother: "If you really want to do this, you're going to have to work really hard. Take advantage of every opportunity, don't give up, and you will find a way." It's a message that informed my life and a message I share with young people around the world, particularly girls in disadvantaged communities. I wish my mother were alive to know how many people have thanked me for teaching them that "because *you* did it, I realized that I could do it too." Today there are many people from all countries who tell me they were inspired to work with animals when they learned about my story.

The 31 women featured in this book are applying GIS technology every day, from scientists to civil engineers, entrepreneurs to urban planners, conservationists to climate experts. They are the strong, passionate women who serve as mentors by inspiring others through their actions. It is my hope that, working together, we can create a critical mass of people who think differently and help to make the world a better place for all people, animals, and the environment.

—Jane Goodall, PhD, DBE
    Founder of the Jane Goodall Institute
    & UN Messenger of Peace

    www.janegoodall.org
    Twitter, Facebook, Instagram: @JaneGoodallInst

PREFACE

In 2018, we at Esri Press thought deeply about a few questions: How could we help bring knowledge and use of GIS to everyone? How could we reach beyond our existing users to people of any age, walk of life, job, or interest? How could we guide young people to GIS as a career and promote the diversity that we see around us every day?

We thought about ourselves, our families, our colleagues, and our idols. And in thinking about their influence on us, we saw an opportunity to influence others. We thought that what would inspire us would inspire many.

In the first volume, *Women and GIS: Mapping Their Stories*, we reached out to 23 amazing women. We saw the vision and the goal of the book clearly—to show young people someone like themselves, so that when they saw them, they might believe that they could do it, too. We got to know these women, tell their stories, understand their challenges, and in the end, make 23 inspiring friends. The reception of that book, from around the world, has been awe-inspiring.

So, we couldn't stop at 23. In our research, we have found an endless supply of extraordinary women using GIS to better the world. In the second volume, *Stars of Spatial Science*, we portrayed 30 more brilliant stars of spatial science, with an equally warm reception.

Realizing that we're still only tapping into the power of people working toward a common goal, in volume 3, *Champions of a Sustainable World*, we bring you equally compelling stories of 31 women+ using GIS for sustainability and growth. And now we know for certain, there is no end in sight to the ranks of remarkable women using GIS to make the world a better place and to sustain it into the future.

—Catherine Ortiz
Manager and publisher, Esri Press

WOMEN AND GIS VOL. 3

Champions of a Sustainable World

PRISCILLA MBAMA ABASI

Making maps and flying drones to save lives

Position

GIS technician
Zipline

Education

MSc in geospatial and mapping science
University of Glasgow, Scotland

BSc in geomatic engineering
Kwame Nkrumah University of Science and Technology in Kumasi, Ashanti, Ghana

*A*s a child, Priscilla Mbama Abasi thought studying science meant becoming a doctor or a nurse, which didn't appeal to her since she hated injections and hospitals. Now she's a GIS technician at Zipline, a logistics company delivering blood, medical products, and vaccines to countries such as Ghana, Rwanda, and the US. Her team works on cutting-edge technology to provide every human on earth fast access to medical supplies and health care. She may not be a doctor, but she still gets to save lives every day. Priscilla says, **"Doing my part to sustain this mission can sometimes be challenging, but that also means I get to problem solve, be constantly on my toes, and be innovative. I love it."**

Priscilla's responsibilities include mapping the delivery maneuvers of Zipline's drones and making sure their flight paths are free of obstacles. She also supports flight and fulfillment operations with geospatial analysis and maps for data visualization. Passionate about making data accessible and visible to all, Priscilla took the initiative to work with the GIS team to create static and web maps as well as analyze geospatial data. This helps teams such as fulfillment operations and customer experience identify patterns and relationships and make informed decisions.

At Zipline, Priscilla helps plan delivery routes and maneuvers for drones delivering medical supplies.

As an only child, Priscilla says she had to be innovative in seemingly small ways, such as creating imaginary friends to play with. This led her to think creatively, often wondering how existing systems could be improved—a skill that has been handy throughout her career. **"My ability to find solutions to solve problems at work is my greatest strength. I am relentless and hardworking, always ensuring that I go over and beyond what is expected of me, no matter how small the task,"** she says, although she acknowledges that "because I always want everything to be done well, I sometimes spend too much time going over and over it. However, I've learned that meeting deadlines is also a part of doing things well, so I try to find the balance between the two."

Making space for women

As one of the few women in GIS in Ghana, Priscilla is passionate about creating space for other women to join the field. She supports African Women in GIS, founded by two young women (Cyhana Williams, one of the founders, was profiled in *Women and GIS, Volume 2*) with the main goal of supporting women who are

Fun fact!

Favorite trip: "I took a trip to South Korea in 2008 as part of a team of six to represent Ghana in the annual International Junior Science Olympiad. That was one of the best times in my life and the first time I saw a movie in 3D. I remember the Korean people I met to be very warm and friendly. There was an instance where we went to a park and a woman asked me if I could take a picture with her daughter because she thought I was so beautiful. Sometimes I wonder if they still have that picture, and I revisit the country through their movies.

"Also, the most memorable and emotional trip I've taken was to the Kigali Genocide Memorial Center in Rwanda. It was really touching to see how a country torn apart by a genocide had forgiven itself and had hope for a prosperous future."

Priscilla hopes to inspire and motivate more women to join the GIS field.

interested in or new to the field and providing a safe space for conversation and discussion. Within this organization, Priscilla manages the social media content team, while also mentoring young women who are interested in the field by providing them with skills to grow their career as well as motivation and encouragement to pursue higher education. "I am passionate about any opportunity to motivate girls to pursue STEM," she says. In fact, because of the lack of female lecturers in her department and in the college of engineering during her schooling, Priscilla hopes to one day become a professor.

Finding her passion

At university, Priscilla decided to pursue biochemistry, with her parents' blessing after they initially objected to her studying science, but she was admitted to the geomatic engineering program instead. Though she didn't like it initially, planning to just become a surveyor because it's what she was trained to do, she says, **"During my master's degree in the UK, I gradually realized the enormous**

Tip!

"Be open minded and willing to learn. Think big! Think about going to space; think about building things no one has seen before."

benefit of geographic information systems in solving problems in my society, and I loved it. I loved giving normal data a face by adding location information."

During her undergraduate degree, though, Priscilla initially struggled at university. "The difference between my results in my first year and at graduation was just the result of hard work," she says. She credits her academic supervisors, John Ayer and Dr. Anthony Arko-Adjei, with showing her what it means to work hard—something that pushes her even now. "The thought of being better and knowing more than I knew yesterday drives me to do better, work hard, and achieve more," she says.

After graduating, Priscilla started as a GIS data manager at a start-up before moving to a GIS analyst position at a tech company that builds geolocation tools for the real estate market. One of the hardest decisions she's had to make, she says, was deciding to move to Zipline, because she was skeptical of its being a start-up. But it worked out for the better: "This has been the most I've grown in my career as a GIS professional," she says. "I have been supported by my team and team leads to learn new skills and improve the skills I already had. As GIS professionals, we tend to be more tool users than tool makers. So to not be stuck in making the type of analysis where I'm clicking buttons and manually moving things around to complete an analysis, I'm learning Python and other programming languages to help automate my work processes. This way I'm not only working hard but smart."

Striving for more

Striving to improve and do better is a constant for Priscilla. Having always admired her mother and aunts for their successful careers and ability to also take care of themselves and their families, she hopes to do the same. "I saw firsthand all the effort my mother put into achieving her goals as well as the hurdles she had to jump,

Priscilla at her master's graduation from the University of Glasgow in 2016.

Tip!

"Don't be afraid to travel any path in STEM that is less traveled by women. STEM will afford you the opportunity to solve problems and impact your society, so embrace it."

Priscilla speaks at the Nima Startup Summit in September 2020. The event was organized by Developers in Vogue in partnership with Ghana Tech Lab.

and that has taught me the values of persistence and self-belief," she says. Priscilla credits her success and career growth to the support of her parents, her loved one, and friends, as well as the Delivery Site Bring-Up and GIS teams at Zipline, who constantly make room for her to learn and grow.

During the coronavirus disease 2019 (COVID-19) pandemic, Priscilla was able to spend time taking a lot of online courses, improving her map-styling skills among other things. And Zipline also made a big impact by delivering COVID-19 supplies in North Carolina and test samples in Ghana. She says, "The pandemic reechoed the popular proverb 'Make hay while the sun shines.' Now I try to live to the fullest and make good use of my time."

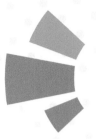

ARIANNA ARMELLI

Taking a business risk to help others avoid risk

Position

Cofounder and CEO
Dorothy

Education

Master's in landscape architecture and regional planning
University of Pennsylvania, Philadelphia

Bachelor's in architecture
New York Institute of Technology, Manhattan, New York

*H*AVE YOU EVER EXPERIENCED a natural disaster such as a flood or wildfire as it was happening or dealt with its aftermath? Those experiences are some of life's most harrowing, and although Arianna Armelli has not directly been caught in a natural disaster, she has been close enough and knows how devastating it is for the people affected.

When New York was hit with Hurricane Sandy in 2012, she says, **"I distinctly remember being able to go to work in Manhattan a week after the storm while friends of mine in the Rockaways were still without power and had lost their childhood homes."** Arianna and her best friend, Maureen, went to the Rockaways a couple of weeks after the storm because a few of their colleagues from school had posted on Facebook that they, and many others, needed help. Armed with a shovel and a sledgehammer, Maureen, who is 5 feet tall, and Arianna, who weighed about 105 pounds at the time, spent the entire day going from house to house slamming through soaked drywall in basements and shoveling debris out of homes. They spoke with some of the families, and every single one of them said they hadn't received aid from the Federal Emergency Management Agency (FEMA) and that their insurance didn't cover floods. Arianna could see that their lives were broken.

Later, Arianna would realize that this wasn't an isolated instance—uninsured home and business owners living and working in areas at risk of flooding or other natural disaster events was a common scenario. Over the years, she would watch countless storms rip through similar areas in other parts of the country and around the world, all with a headline alluding to ineffective risk maps and massive uninsured communities devastated after a storm. Arianna founded Dorothy, a tech company that collects and transforms data from hundreds of sources, allowing insurers and insurance policy holders to understand the real risk of their policies. "I founded Dorothy because I was passionate about finding a solution to what I recognized as a very serious problem," she says.

Arianna, *left*, and her friend Maureen help homeowners in the Rockaways after Hurricane Sandy in 2012.

From play builder to architectural professional

Arianna's mother, Rochelle Shapiro, always said that Arianna would be an architect because she used to spend hours by herself building cities with Legos. "My mom filled an entire bookcase with architecture books before I even got into college because 'I was going to need it someday.' Now she sends me articles from *Forbes* and *Entrepreneur* magazine any time they feature a [strong] woman," she says.

Even more than helping Arianna recognize one of her early interests, her mother had the biggest positive influence on the woman Arianna is today. "It's not so much anything my mom has said but who she is that has inspired me," Arianna says. "My mother has always been unashamedly herself in the face of people who wanted her to be something else. Through the 31 years of being her daughter, I have never once seen her cave under pressure or change who she is to accommodate the comfort of a weak person. That single trait has taught me to be strong in a moment of confrontation and patient in the face of difference."

Arianna grew up in a small apartment in the Bronx with a working-class family that was basically happy and modest. In high

Fun fact!

"I have always been conscious of my fitness and usually maintained my well-being through active workouts, either cycling or weightlifting. During the pandemic, I found that my regular routine was no longer possible, so I took up running, yoga, and meditation. I feel like I'm a bit late to the party, but it changed my life."

school, she started exploring the architecture field by taking Auto-CAD classes and getting involved in housing design projects. She continued that interest into college at the New York Institute of Technology in Manhattan, New York, where she studied urban design. She became fascinated by city planning because city design was the outcome of politics, economic interests, and industry forces. Her focus was primarily on large-scale designs located in or near flood zones.

Her parents worked incredibly hard to provide her with an education that they didn't have the opportunity to pursue while they were growing up. They paid for her first year of private college until they could no longer afford it. Arianna paid for the rest of her education herself and built a decent savings account based on her architecture income. Her parents taught her how to value money and the hard work that came with it. "I was taught that there's no reward without risk, and if you have the luxury to coddle failure, you'll never truly feel the gift of success," she says.

College is where she met all her closest friends; it is also where she met one of her most influential teachers, Giovanni Santamaria. She found herself involved in Giovanni's urban design thesis, which was a year-long study focused on the rejuvenation of postindustrial landscapes. The thesis involved intense research starting with geopolitical forces that trickle down into the economies of cities and how those geopolicies influence design and infrastructure. "Giovanni taught me to question the obvious and to solve large problems through analytical approaches," she says. "He has been a mentor to me for 10 years now, and whenever I need his advice or input, he never lets me down."

Arianna's father died unexpectedly the year she graduated from college, and his death harmed her family financially. At the time, it was important for Arianna to build stability with her career, so she lived at home and helped where she could.

Arianna and her father, Joseph Armelli, at her graduation from New York Institute of Technology in Manhattan.

Top: Arianna with Prague behind her during a meaningful trip to Eastern Europe.

Bottom: Arianna's journey started in Budapest, Hungary, and ended in Warsaw, Poland.

During her last year of college, she began her professional career as an intern at a small architecture firm in midtown Manhattan that primarily focused on mixed-use residential and commercial developments along waterfronts. The position she was offered after graduation came from part luck and part personal ambition. "I was 22 and wanted to lead a new 1.6 million-square-foot project they just got in the door that would take up an entire city block in Queens. I was young and eager and in over my head," Arianna admits. She spent the next six years developing that project from sketch to structure, working with mechanical, structural, and civil engineers, contractors, and the city government. She remembers, "The entire project felt like a never-ending problem that had to be solved, whether it was zoning changes, building department approval, community board input, FAA restrictions, underground parking structures that were 30 feet beneath the water table—**every day was a newer, harder problem that needed tending to. In six years, I learned how to solve massive problems on the daily, but most importantly I learned how to build nothing into something.**"

Arianna and her mother, Rochelle Shapiro, at Arianna's graduation from the University of Pennsylvania.

Grad school and Arianna's first company

Arianna left her architecture job in June 2016 and began grad school at the University of Pennsylvania in August. Her program focused on large-scale regional planning with an emphasis on sustainable design approaches that integrate landscapes. She decided to attend grad school for a couple of reasons: (1) she already knew she wanted to start her own business and wanted to build a more diverse network of colleagues from other professions, and (2) she knew she needed to know more about business and engineering and made a concerted effort to fill her electives with classes outside the design school, including as many engineering and business classes as her schedule allowed. When she came up with the initial idea for Dorothy, she was able to use all the university's entrepreneurial resources, which included incubation hubs, start-up grants, and co-op working space at the Pennovation Center.

Just before she started grad school, Arianna founded her first company, Nativah Chaya. She initially started it after completing a 50-mile bike ride through the five boroughs in New York City, which she now calls her bridges tour. "The morning of the ride, I woke up with the intention of biking to the High Bridge over the Harlem River—it had just reopened as a pedestrian overpass after some 30 years of being closed to the public," she recalls. "It was such a beautiful day that once I got there, I felt I should explore more. That day I cycled past every bridge connecting to Manhattan. Chaya is my middle name, which means life in Hebrew. *Nativah Chaya* means paths of life. I had learned the figure-ground method of mapping in undergrad and utilized GIS systems and code to allow individuals who wanted to commemorate a special journey or achievement by syncing their fitness app with my mapping designs. While my intention for Nativah Chaya was to create a unique project, the company would later support most of my expenses during grad school and provide income during the early stages of a boot-strapped Dorothy."

Arianna memorialized her 50-mile bike tour through the five boroughs of New York City, which became her inspiration to start her first company, Nativah Chaya.

Jumping into innovation and the unknown

In 2017 after several hurricanes landed in Texas and Puerto Rico, Arianna read in the paper that "about 80 percent of Hurricane Harvey victims" did not have flood insurance and faced huge bills (*Associated Press*, August 29, 2017). Many of those homes fell outside the federally backed flood zones, according to the story that appeared in *USA Today*. At this point, she realized urban design was a long-term solution to a short-term problem and that she did not want to spend her career working on one project that would affect only hundreds of people. She wanted to come up with a solution that could affect hundreds of thousands of people in real time. So she shifted her focus to technology and began researching how FEMA flood maps were created and why their technology was providing such conservative risk assessment.

Arianna knew that solving problems at someone else's company was not going to be enough. She wanted to start her own business focused on solving problems that make a difference in people's lives.

Even though she knew that she wanted to start something on her own, Arianna still feared the lack of stability that she would otherwise get from a steady job and income. "The hardest choice, which ultimately became my sacrifice, came when I decided to jump, not just talk about it," she says. "I financed Dorothy with my entire savings and relied on my own confidence in myself to make it happen. **Entrepreneurship is certainly a journey of high highs and lower lows, yet persistence and patience have pushed this company forward, and the feelings of accomplishment after doubt are incomparable to anything I've ever experienced."**

Dorothy's core technology focuses on producing more accurate predictive analysis for natural disasters and streamlining the coverage process for disaster insurance. For example, the technology that Dorothy uses mapped flood damage from Hurricane Harvey 60 percent more accurately than standard flood maps, she says on riskandinsurance.com. Considering the increase in climate-related storms, as well as their severity, Arianna and her company use historical reference as a data point but say that it should not be the sole source of risk-based assessment.

If Arianna had listened to all the people in her life who told her no or said that it's going to be too hard, she wouldn't be where she is today, she says. Today, she is the CEO of a funded company with what she calls a phenomenal team that solves problems they care about. Yet she appreciates the educators who discouraged her because it made her work a little harder to prove them wrong. She knows how hard she is willing to persevere. "I never quit," she says. "Success is very simple to me. It is the ability to achieve everything I want with my career while providing for the people I love along the way."

As Arianna pursues a goal she is passionate about, she deals with all the feelings that come with building something from nothing. It's a lot of responsibility, and she constantly wrestles with the fear of failure, she says. But she learned valuable financial lessons

"Before the pandemic, I was 110 percent focused on my job. It was the only part of my life I catered to with unlimited devotion. Through COVID-19, I learned that type of lifestyle is not sustainable and that the most important part of life is living it. I have taken the time to slow down and appreciate the journey while making a concerted effort to spend time with the people I care about. Don't get me wrong–building my business is still my number one priority, but it is also possible for me to spend time in the present without emotionally paralyzing myself with thoughts of work that still needs to be done. Meditation helped with this."

at home and applies them daily. Plus, she possesses an unwavering ability to persevere.

"I would encourage young women to pursue a career in STEM simply because there are limitless opportunities for professional autonomy and financial independence," she says. **"If you are like me and crave the freedom to explore a path of the unknown, aka entrepreneurship, a career in STEM will foster the technical foundation to achieve those goals. I feel no regret for leaving what could have been a very successful career to focus on building something myself and would urge others wrestling with the same thoughts and feelings to do the same."**

MARYGRACE BALINOS

From imagined cities to real solutions

Position

Freelance GIS specialist and entrepreneur
Valdivia, Chile

Education

Postgraduate diploma, energy efficiency and environmental quality in construction
Universidad Austral de Chile, Valdivia

MS, regional development planning and management
Universidad Austral de Chile, Valdivia, and TU Dortmund University, Germany

BS, civil engineering
University of the Philippines, Los Baños

WHEN MARYGRACE BALINOS wanted to join an art club in high school, she had to demonstrate her talent by drawing her vision of the future. She still remembers "the spontaneity of the colors and lines of the skyscrapers" that she drew in her imaginary city. For a while, she thought she might pursue a career in the arts, but encouraged by her mother, who was a college math instructor, she chose instead to become a civil engineer. She discovered GIS while studying at the University of the Philippines, and today, among other projects, she is working on a smart city system in Valdivia, Chile, applying her vision and her skills to transform a real city into a city of the future.

In 2018, with a small team, Marygrace submitted an innovative idea for the Smart City challenge organized by InnovING 2030. Her proposal for an open-source smart system to monitor the municipal waste collection in the City of Valdivia was selected, and the team is currently finishing the mobile and web app prototype, Ciudad Limpia. In addition to monitoring, the system is designed to promote a circular economy and the reporting of unregulated dumping. "One of the social implications of the project that we envisioned is bridging the communication between the municipality and the community," Marygrace says. "The project development coincided with the social transformation that is happening in Chile (university strikes, citywide demonstrations, social unrest) and very much evident in Valdivia, in addition to the current health crisis—the COVID-19 pandemic. Thus, finishing the prototype has been a challenge, too." Ciudad Limpia was expected to launch in 2021.

Discovering GIS

It's a long way from the Philippines, where Marygrace was born, to Chile, where she works as a freelance GIS specialist, currently making maps for a proposed national park management plan in Northern Chilean Patagonia. Her journey began during her undergraduate years, when she worked as a student assistant at the GIS-IP Laboratory at the International Rice Research Institute in Los Baños, Philippines. "My first task was to digitize paper maps using a digitizing board and ArcInfo software," she recalls. "It was a new technology for me, and that started my interest. By the way, one of my direct supervisors was a woman."

Marygrace notes that she was surrounded by women in her formative years and that almost all her immediate supervisors when she started working were women. And, she says, "My mom is the woman I admire the most. She was hardworking, very good at her craft, and sacrificed a lot for us, her children. She always told us the importance of education, being kind to others, and to always have faith in God." Apart from her family, Marygrace credits her "closest and dearest friends in college" for supporting her. "We did a lot of fun and interesting activities and group studying together," she says, "and I love and appreciate them very much."

After graduating and qualifying as a licensed civil engineer, Marygrace worked for a private construction company in the Philippines, where she applied her GIS expertise and spatial thinking to real-world problems for the first time. "One of the challenges we faced was coordinating with the site and planning the optimal route for the trucks that carry the construction materials from the

Marygrace, *right*, and her mother at the oath-taking ceremony for new civil engineers in Manila in 2003.

warehouse to different construction sites," she says. "It was a small thing, but I was proud of it, since most of what I knew was theoretical, and at that moment, it was a real situation. Acting quickly to resolve issues, taking into account the security of our coworkers, is very important in that line of work."

A growing interest in urban planning led to a postgraduate diploma, and then what she describes as a wonderful opportunity. On her second try ("Perseverance is one of my strongest traits," she says), Marygrace was awarded a scholarship to an international master's program in regional development planning and management, known as the SPRING—Spatial Planning for Regions in Growing Economies—programme. She completed her first year of study at Technische Universität Dortmund, Germany, and the second at a partner institution, Universidad Austral de Chile.

A difficult decision

Although she had been living away from home since college, the decision to leave the Philippines was difficult and came at significant personal cost. "The hardest choice that I had to make was to leave my family to study abroad," she says. "I asked Mom for some guidance, and she was convinced that I should continue that path. The last time I hugged her was when she was sending me off at the airport. After about six months, my mom passed away due to an accident."

The sacrifices Marygrace made to further her education were real, but so too were the professional opportunities and avenues for entrepreneurship that she has encountered in Chile. After finishing her postgraduate degree, she continued working in Chile as a civil engineer in the public sector and as an academic assistant, coteaching the GIS coursework in the master's program she graduated from. Then she spent eight years working as a GIS assistant and later as a spatial planning technical officer for an international NGO focused on environmental conservation.

Marygrace's family, with her mother in red, sending her off at the Ninoy Aquino International Airport in 2006 to study in Dortmund, Germany, for her master's degree.

Marygrace at Machu Picchu,
Peru, in 2008.

Fun fact!

Favorite trip: "Traveling to Cusco and Machu
Picchu in Peru. In the engineering library of
my college, I used to go and skim through old
civilization atlases and admire the structures
built during that period, and one of them is
Machu Picchu of the Inca Empire. Being there
was just so surreal—we waited for sunrise,
and as the fog subsided, the city emerged.
Breathtaking!"

Marygrace in Concepcion, Chile, in September 2019, inviting Chile student members of the Institute of Electrical and Electronics Engineers to the 2020 Latin American Geoscience & Remote Sensing Society and International Society for Photogrammetry and Remote Sensing conference.

DondeLaViste?

While working for the NGO, Marygrace came up with an idea: Why not create an app so that members of the community could use their smartphones to report their sightings of various species and, in this way, register them for geotagging? The proposal was funded in 2016, focusing on marine fauna, and, Marygrace says, "I was really happy that the idea became a reality. The mobile and web app is called DondeLaViste?, a collaboration between the NGO and the academic community. At present, the app has about 1,063 installations and more than 500 sightings registered, and it also has the support of the Ministry of Environment." The community sightings, once submitted, are validated by species experts, and beyond geotagging, the project aims to protect the species through environmental education and sustainable special-interest tourism.

A brave move

In late 2019, Marygrace took a leap of faith and decided to become a freelancer. "It was a brave move on my part," she says, "and it was not easy, especially now with the pandemic, but the journey was interesting." She upgraded her skills by taking online courses in

Fun fact!

"I make jewelry as a hobby and joined an exhibit showcasing our jewelries with recycled metals. I took a wood carpentry course and made my own shelf without using nails. I also took sculpting classes and created a foot-tall clay sculpture of a penguin."

R programming and teaching herself Python and CSS. "There were a lot of firsts, like applying for a start-up and participating in an international innovation technology competition co-organized by the GEO-Land Degradation Neutrality (GEO-LDN) initiative and the United Nations Convention to Combat Desertification with the support of my husband, who is also a partner in our small company," she says. She was thrilled when their proposal, Land Use Planning assistant (LUPa), was selected as one of the competition's semifinalists.

As she looks toward the future, Marygrace is motivated by the motto that has inspired her personal and professional journey thus far: "I can do it!" It's an attitude that she would like to pass on to other young women considering a career in STEM. **"STEM is as fun as any other fields," she says. "Engineering is an area of study where we can imagine, create, and invent a lot of things that not only help in our everyday lives but may also have a great and positive impact in our society.** We have equal skills to men's. It is only a matter of perspective and training. Believe in yourself! If we want to do it, we can achieve it."

"I can do it!" Marygrace participating in an interschool gymnastics competition in the Philippines in 1989.

FIONA BECKER

Blending the beauty of science and art

Position

Conservation Information manager
The Nature Conservancy

Education

MS in environmental education
The Audubon Expedition Institute, Lesley University, Cambridge, Massachusetts

BS in biology with a minor in music
Millikin University, Decatur, Illinois

CONSERVATION OF THE NATURAL WORLD is Fiona Becker's passion, and she has the great joy to work in a field that she is passionate about. Her interest in the natural world started early: her mother tells the story of finding Fiona out in the backyard as a toddler, unconcerned that she was covered with ants, just fascinated as they crawled all over her.

Her family lived in Chicago, Illinois, and spent weekends throughout each year taking camping trips. Every summer her parents took Fiona and her three sisters on two-week adventures in their 15-foot travel trailer across the US and Canada. As they traveled, Fiona endlessly pored over maps, figuring out distances between their destinations, following the roads they were on, and helping find campgrounds. **"My love of maps has followed me into adulthood,"** she says. **"My husband often jokes that if I'm cranky when we travel, he just needs to hand me a map to cheer me up."**

As a child, Fiona wanted to grow up to be a ballerina or a firefighter. She loved books and learning and, with only a few exceptions, was blessed with teachers who were excellent educators. She recalls a physics teacher who translated his subject into something that was accessible and exciting to high schoolers and a choir director who gave his all to his students, helping them become better musicians and opening their eyes to broader experiences through performing. His guidance helped Fiona choose a college that would help nurture her love of music but was also diverse enough to foster her interests in the sciences.

Of course, there were a few teachers and counselors who discouraged her as well. She recalls losing a fight to keep Physics 2 and choir on her high school schedule in her senior year, because both were electives and she had to choose her love of music over her love of science that year. She also remembers an advanced calculus teacher who fawned over students who easily grasped difficult concepts but mocked those who struggled to understand.

In the fall of 1991, Fiona began as an undergraduate at Millikin University, a small university in Decatur in central Illinois. She started as a double major in biology/pre-physical therapy and vocal music, two majors without much overlap in coursework. Her first two years were joyful and exciting, but she finally decided to choose a single major: biology, with a minor in music.

Her early years of ease with education altered dramatically as an undergrad; she struggled to learn how to study effectively and had to work harder to make good grades. As she put in observation hours in physical therapy, she realized that it was not a good fit for

Fiona works in conservation and GIS for The Nature Conservancy.

In this childhood photo at the top of Pikes Peak in Colorado, Fiona is second from the right, standing on the bench. Her family's travels inspired Fiona's current bucket list to visit every US state (currently she has visited 49) and world continent (only three so far).

her future career. Luckily, it was simple to shift the focus within her biology major. She was just starting to take higher level biology courses that helped her figure out where she wanted to continue—entomology, botany, animal behavior, all ultimately leading to conservation.

Fiona credits Dr. Marianne Robertson, biology professor for her junior and senior undergraduate years, with being instrumental in helping her focus her passion and start her on the path to her future career. Fiona says, "Marianne's sheer energy and exuberance for her chosen subject and how she shared that passion and knowledge with her students motivated me to find a career that I could feel passionate about."

An education while on expeditions

After graduation in 1995, Fiona began a master of science in environmental education through the Audubon Expedition Institute, an accredited degree program available through Lesley University in Cambridge, Massachusetts. Her campus stretched across North America. As Fiona describes it, "For my first three semesters, I was part of a group of 18 to 25 students and four instructors traveling through a bioregion. We traveled in a modified school bus that housed all our belongings, gear, and a library. We camped and lived outdoors almost entirely for each of our 13-week semesters. Each semester had several long hiking or canoeing trips, together adding up to a month in the backcountry." As they traveled, the students were taught by instructors and educational resources from across a wide spectrum of environmental issues, from conservative to radical. These resources included people such as author Howard Zinn; Senator Joe Biden; a founder of Earth First; sustainable foresters; employees of an international timber harvester; Isabelle Ides, who was one of the last native speakers of the Makah tribe on the northwest coast, and many more.

By design and necessity, this educational program was low tech. Fiona had to relearn how to research, write, and edit without the aid of a computer for all but the final drafts of her papers. She learned how to develop and teach seminars and how to live and work as part of a small community led by consensus decision-making. In her final semester, she completed her graduate degree with an internship at Openlands, a conservation organization in Chicago. She didn't realize it at the time, but with Openlands, she was making her first foray into using a GIS as she assisted with the photographic survey of what would eventually become Midewin National Tallgrass Prairie, a large conservation project in the US Midwest.

After graduate school, Fiona embarked on a series of seasonal positions while sending out what felt like hundreds of applications and resumes for jobs in conservation. Her first seasonal position was with the City of Chicago Department of the Environment, teaching public school kids how to landscape and helping them plan and implement a landscape design for their school.

Fiona worked with junior high students at Pasteur Elementary School through Mayor Richard M. Daley's Summer Environmental Schools Initiative. This Chicago public schools program taught students about nature and environmental issues and helped students design and install a landscaping project around their schools.

Shown here aboard the *Van Elliot*, a 70-foot cod pot boat, Fiona was a Fisheries Service observer on commercial fishing vessels in the Gulf of Alaska and Bering Sea. While she was onshore in Alaska, she saw moose and bears, and the northern lights for the first time.

Next, she answered an environmental job ad that asked, "Have a biology degree? Want to see Alaska?" That sent her on an adventure as a National Marine Fisheries Service observer on commercial fishing boats in the Bering Sea and the Gulf of Alaska. After a three-week course learning how to identify and sample North Pacific fish and shellfish, collect data, and survive and be safe in extreme cold weather and water, she spent the next six months onboard long-liners, trawlers, and pot boats, counting fish and crabs. Fiona recalls:

It was an epic experience, somehow both thrilling and monotonous. I was always the only woman on boats that were out at sea for anywhere from a week to a month at a time. I was generally the most disliked person on the boat, since I made a difficult job even harder as I sampled their catch, and the reports I made back to the Fisheries Service could mean that a fishing season might be closed early. **During my training, I was taught secret codes that could be sent with my weekly radio reports, to let them know if I was in danger or being forced to** report false numbers by the skipper or crew. Luckily, my crews took my desire to follow the rules with only some minor grumbling, and I never had to use those codes. I got to experience 40-knot winds and 70-foot swells in bad storms. I watched as orcas snacked on most of the catch brought up on a long line before the skipper called for the line to be marked and dropped until the orcas got bored and swam away. I saw an enormous halibut get pulled up in a cod pot and estimated its weight at over 200 pounds; the fishermen and I were all happy to send it safely back into the Gulf to live longer and have many more baby halibut. I saw a fisherman nearly get swept overboard. I hitched my first ride ever in Dutch Harbor, in the Aleutian Islands, to get from the docks to the grocery store. And I counted lots and lots of fish and wrote endless reports.

Eagerness and good references win the job

After her second, three-month contract with the Fisheries Service was completed, Fiona returned home to Chicago. She interviewed for a summer job as a field assistant with the McHenry County Conservation District in northern Illinois. Fiona vividly remembers this job interview and being told that, with her education, she was overqualified for the low-paying position. It became clear that she lacked conservation field experience, and the only way to get it was to work an entry-level, seasonal job. To accept the position, Fiona would have to let go of the multiple, part-time jobs she had juggled in previous years. These jobs had paid more but weren't helping her advance her career. The decision to take the temporary drop in income was a crucial step in her career path.

Fiona recognizes that her privilege and support network made it possible to take that financial risk.

Fiona worked hard that summer doing prairie and wetland restoration. She says, **"I learned that habitat restoration often requires the use of power tools, and that chainsaws are very heavy and oddly satisfying to use. I learned to identify native and nonnative trees and plants. I got to do stream surveys and identify freshwater fish. And in the end, taking this physically demanding, low-paying position was the best possible decision, as it led me to a job with The Nature Conservancy."**

During her graduate studies, Fiona had heard about The Nature Conservancy (TNC) and had been applying for jobs with them since graduation. As her seasonal position with McHenry County ended, nearly an identical position opened up with TNC as a grant-funded, two-year post in northwest Indiana. The interviews went well, and Fiona found out she was one of two people being considered for the position. If she didn't get the job with The Nature Conservancy, she would need to go back to Alaska and the fishing boats. The fisheries contractor had left a voicemail asking

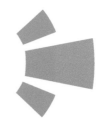

Fiona adds chainsaw art to the stump of a tree she cut down while thinning cedars in glade habitat in southern Indiana. Cedar trees can shade out other native prairie grasses and flowers in these natural forest openings. Thinning the trees mechanically or with prescribed fire provides space and light for other native species to flourish.

Fiona, here with TNC preserve steward Bruce Shepard, preparing for a prescribed fire at Gravel Hill Prairies, a small nature preserve in central Indiana. Fiona trained to be a wildland firefighter and has been on dozens of burns as part of TNC's prescribed fire crew.

when they could schedule her return. Fiona recognized that she needed to act.

Although it was a risk, Fiona reached out to her references, asking them to call the interviewer at TNC. The risk worked out: her eagerness and good references tipped the scales in her favor, and she got the job as a restoration specialist at TNC.

Her first year, 1998, with TNC was like her restoration work the previous summer. She used chainsaws and brush cutters to help restore prairie and wetland habitat in the dunes and swales in northwest Indiana. She realized her childhood dream of being a firefighter as she trained to be a wildland firefighter as part of TNC's prescribed fire crew.

In addition to habitat restoration work, her manager discovered that she was also good with computers, so she began to assist with office work and writing reports. At the time, the office had been working with a nearby university to create a GIS for a site conservation plan of the project area. When the data was delivered, it needed to be shaped and organized for the final plan. Fiona's manager handed her a disk full of shapefiles and a set of ArcView 3 software disks and asked her to figure it out and make some good maps. With some struggle and stubbornness, she did.

Since then, Fiona has made more maps than she can count. In 2000, she shifted to a permanent position at TNC's Indiana field office in Indianapolis. She has worn several hats at the Conservancy, including site conservation planner and information technology manager, all the while fulfilling the need for a GIS person for the Indiana chapter. She took classes and grew its GIS capabilities. When she moved to Indianapolis, she and her manager were the only TNC staff using GIS in Indiana. "I helped change that," she says, "and now basic GIS skills are a requirement for all our 25-plus stewardship staff. I am incredibly proud to have helped train these preserve stewards in using GIS to track and manage their fieldwork. With the help of the stewards and GIS experts at TNC, and with the improvements in technology, we have shifted from the paper forms we used more than a decade ago to using smart devices out in the field to collect our field activities. With all our stewardship and monitoring activities in our GIS, we are better able to pull data for analysis, measure success with our restoration methods, and report our successes or failures. **Plus, it's always exciting for me to have a 'magical map' in my pocket at all times since I can load any TNC preserve in Indiana onto my phone while I'm on the road."**

In addition to conservation planning and analysis, Fiona takes the science that drives TNC's conservation work and turns it into ArcGIS® StoryMaps stories that the Conservancy can share with the world about the amazing work they do. Often, she helps edit presentations and posters, providing maps and editing content to communicate best with the intended audience. Fiona helps the partners and donors for TNC understand where conservation work is done and how much TNC protects with their help. Other maps help staff make strategic decisions on where their work will be more successful and where to focus their efforts. **"It's a bit of a dance to provide what my colleagues want with what will actually** tell the story they need told," she says. "These maps typically end up in funding and grant applications and reports, newsletters, our website, preserve signage, and social media." She has also served as a member of the Indiana GIS Council through most of the last two decades.

When Fiona moved from a field position to an office position, she missed being outdoors during her workdays. To counteract that loss, she began creating her own native gardens at home, in a residential neighborhood just a mile and a half from the center of downtown Indianapolis. After a few years, her front yard prairie planting grew in stature and beauty. Its healthy growth, however, drew the attention of the city code inspectors, who labeled her yard an environmental nuisance. At that time, the mayor of Indianapolis was promoting rain gardens and native plants to help make Indianapolis a more sustainable city, yet the municipal code still noted that any vegetation over 12 inches tall was considered a weed. Fiona embarked on a crusade to help change that city code. "It was in my favor that the Mayor's Office of Sustainability was also working on the same thing. They enlisted me to testify at city council meetings to amend the code," she says. "My yard became an example of how to design a native planting and was the first garden enrolled in the Indianapolis Native Planting and Rain Garden Registry. This experience got me motivated to help other folks design and create their own urban native plant habitats. Additionally, I participated in the design process and maintenance of the 10,000 square feet of native plantings around The Nature Conservancy's main office in Indiana, a LEED Platinum certified building near downtown Indianapolis." Fiona has also served on the board of the Indiana Native Plant Society.

Fiona believes she can figure out almost anything with time and stubbornness, but this attitude has also proved to be a weakness. "Knowing when to call in an expert was something that I needed to learn to do," she says. "Knowing that it is a better use of

Fiona sits among native Indiana plants that she helped plant around the Efroymson Conservation Center, TNC's Indiana field office, in downtown Indianapolis.

time and resources to have a contractor do a complicated analysis was a hard lesson for me to learn. More importantly, having the confidence to tell my managers and colleagues that I was not the right person for their GIS request was a difficult obstacle for me to overcome." For Fiona, success is earning the respect of her colleagues and loved ones for the work she does both in her professional and personal life.

When she was a recent college graduate, it was hard for Fiona to imagine what she would end up doing with her biology and environmental education degrees. "It is amazing to me that I was able to find and ultimately create a job for myself that draws so deeply on both of those degrees, and also allows me to use my artistic skills that I wasn't willing to give up as I studied other subjects," she says. Fiona finds beauty in GIS. **"I feel like we are often told that there is a stark divide between the love of arts and the potential for career success. You can love the sciences and technology and love art. You can combine them to find a successful career. There doesn't need to be a divide. You can have both.**

"My work at TNC is primarily with scientific facts and figures, but with artistic design, the power of that science sends a much stronger message. I have also kept music in my life by dancing, playing music, and singing with family and friends and for lots of wedding ceremonies. I have even gotten to sing a few times for TNC's Facebook page."

As one of her milestones at TNC, Fiona was asked to join the inaugural Geospatial Leadership Council in 2019 to guide the more than 1,300 GIS professionals across the organization. The Council is set up to design and promote a geospatial vision that supports conservation efforts locally and around the world. As a founding member of the Council, Fiona's breadth and depth of GIS experience has helped direct the many ways in which the community contributes to the organization's conservation success.

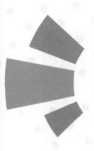

MAGGIE CAWLEY

Traveling the open road for open data

LWAYS KEEN TO GROW AND LEARN, Maggie Cawley is passionate about the power of collaboration. Throughout her travels, such as her work creating maps in southern Africa and Mauritius, she has seen firsthand that working with others can make a positive impact on a community and the world. Currently, she is motivated by working alongside other determined and dynamic people to expand the impact of OpenStreetMap (OSM), a collaborative project to create a free editable map of the world. The geodata underlying OpenStreetMap, a map created by millions of people around the world, is considered the primary output of the project, showing how the efforts of many can be greater than the efforts of individuals. In her role at the nonprofit organization OpenStreetMap US, Maggie creates space for collaboration, amplifies the voices and innovations of the OpenStreetMap community, and helps build the broadest open-source map of the world.

Bitten by the travel bug

At an early age, Maggie developed a passion for the environment and curiosity for the world. Raised in a solar house built by her parents, she was taught resourcefulness, conservation, and respect for the environment and the earth. Her interest in places beyond her own backyard grew from

Position

Executive director
OpenStreetMap US

Education

MURP (master's of urban and regional planning) in urban planning
Virginia Commonwealth University

BA in international studies in world politics and diplomacy
University of Richmond, Virginia

As the executive director of OpenStreetMap US, Maggie speaks at the State of the Map US conference in 2019.

family travel tales and the pages and pictures in *National Geographic* magazine. With each issue, the number of maps pasted on the walls grew. Growing up, she aspired to become a photographer for *National Geographic* or the first female president of the United States. In high school, Maggie took her first trip to Europe with her German class. Over 10 days, the class experienced castles in Bavaria; culture in Munich, Germany; the mountains of Innsbruck, Austria; and the gondolas of Venice, Italy. Maggie says, "It was incredible and life changing, and I was definitely bitten by the travel bug." Travel would become a lifelong pursuit.

Maggie spent her undergraduate years at the University of Richmond in Virginia, where she was a Bonner Scholar. Through the Bonner Scholars Program, she received a scholarship in exchange for meaningful, community-based weekly and summer volunteer service. Every summer, the program offered an alternative service trip, and Maggie used the opportunity to work alongside Hungarian students, teach English and support flood remediation, help build a monastic labyrinth in Ireland, and travel with a group from Tennessee to work in and explore national parks in the American West.

In college, Maggie considered joining the Peace Corps for a more immersive experience somewhere in the world but upon graduation decided to apply for a position with the International YMCA in Germany to make use of her years of German studies. After sending the application and the fee, Maggie discovered that the person she was interacting with for this opportunity was a fraud. Devastated and left with little of her savings, she moved in with a friend and found a job with a temp agency. It was a disappointing experience, and one that changed the trajectory of her life.

One of her first temporary jobs was in the office of a real estate development firm. Maggie started as an assistant but was brought on full time after only a few weeks. As the only staff member, she quickly had many new responsibilities. She managed the monthly contractor payments, took the site development photographs, and provided press and communications support for the transformation of a historic building into a luxury hotel. After her first year, she wanted to learn more about the field and started attending Virginia Commonwealth University (VCU) at night for a certificate in adaptive reuse, which evolved into a master's in urban planning over four years. Maggie's first job taught her about real estate development and the industry's culture. She was often perceived as the assistant who was there to get the coffee or the woman to catcall as she took site photos. Her hopes of becoming a construction manager were dashed when her bosses saw her as better suited to selling condos. The silver lining was that the atmosphere motivated her to finish her master's degree and pursue a new path. While at VCU, Maggie worked for an environmental consulting firm in Maryland where she worked on a water resource element as part of a county comprehensive plan. Her work doubled as her thesis, and upon graduation, she was offered a full-time job at the firm. She said goodbye to Richmond and headed north to Baltimore to begin the next phase of her career.

Fun fact!

When working and traveling in Africa in 2016, Maggie showed photos of her experiences to a Danish musician who hired her to be a photographer as his band toured. She met other Danish and Ghanaian musicians and joined a new Afrobeat band called African Connection that toured southern Ghana in 2017. Maggie started on cowbell, but once they learned she had studied piano since age seven, she became the keyboardist. Maggie says it was an amazing experience and hopes she can play with them again someday.

During her travels, Maggie became a founding member of the band African Connection and was able to tour Ghana in 2017.

Maps were in high demand at the consulting firm, but Maggie's GIS knowledge was minimal. A colleague took her under his wing and began to pass overflow projects her way. She said yes to every opportunity to learn on the job. A year later, she enrolled in a GIS certificate course at Johns Hopkins University, and by her third year, she was a GIS analyst not only for the Annapolis office, but also for the Washington, DC, team. Maggie also started traveling to conduct on-site work for the firm and was the first to raise her hand when the chance for positions outside Maryland arose. She learned how to delineate wetlands and take bathymetric measurements. She traveled to Pascagoula, Mississippi, to support the BP Deepwater Horizon oil spill recovery effort in 2010. But time and again, she was overlooked for international projects. She eventually learned that her GIS skills were in such high demand in both stateside offices that the firm could not afford to send her abroad. Frustrated and discouraged, Maggie started saving and planning for her own trip around the world and eventually requested a summer sabbatical.

That summer she worked with a team of high school students to build trails in Baltimore through the Student Conservation Association. She knew this wasn't her future career path, but she says, "Working with those kids got me out of a planning office and back into what felt more like the real world—that place where I was motivated to try and make a difference." At the end of the summer, the consulting firm offered her more money and a position in management to return but she declined. Maggie wanted to see the world, and there was no going back. She says, **"The choice to leave wasn't easy. I had everything we are taught to aspire to: a '9 to 5,' health insurance, a monthly paycheck, and a corporate ladder to climb. But I chose to follow my heart and live my dream of traveling the world and helping to make it a better place."**

Tip!

"GIS is a tool that can be used no matter what career path you choose, so don't hesitate to change directions because you never truly know where the next turn will lead."

Starting a GIS business

In 2013, Maggie founded her own business, Boomerang Geospatial, to bring GIS into the hands of organizations working to positively impact communities. She also discovered open-source software and open data, which made geospatial tools even more accessible. It was a slow first year, but she had a few early wins—she taught GIS to high school students for a Parks and People summer program and was hired for a small planning project for a nonprofit organization in Baltimore. That fall, she sold many of her belongings, put her GPS and laptop into a backpack, and took Boomerang for the trip she had been planning around the world. She didn't know it yet, but she would not have a permanent address again for six years.

Her first destination was South Africa, and as fate would have it, her first week there she was hosted by someone looking for a few more passengers to join a three-month overland camping trip through 23 national parks in South Africa, Mozambique, Malawi, Zambia, Botswana, and Namibia. It was an invitation she could not refuse.

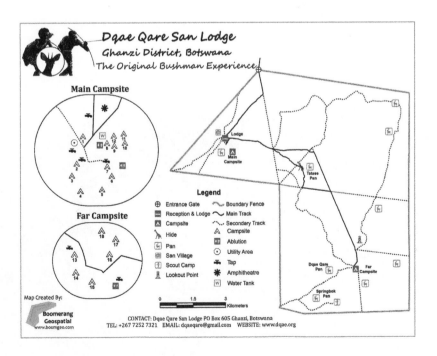

The first map Maggie created for the Naro San reserve.

A highlight of that trip was in Botswana at a game reserve belonging to the indigenous Naro San in the Kalahari Desert. When Maggie arrived, she did not see an accurate map of the campsite or property, so she offered her services in exchange for internet access and a few nights in a real bed. Over the next week, her team of three travelers navigated the entire reserve (more than 150 kilometers) and mapped every road, pan, campsite, water tap, viewpoint, and amenity they could find. She fell in love with the place and the community and vowed to someday return.

Those three months traveling across Africa in a Land Rover, taking every day as it came, were life changing and solidified her belief that communities need access to data—free, open data. Along the way, so many places had either limited or no data about

Maggie in the Namib Desert on her first three-month tour of southern Africa.

their communities or nature reserves. She wondered, how do we preserve something that we don't know is there? Maggie says, "I experienced some of the most incredibly wild, authentic, and beautiful places I had ever imagined. I also witnessed extreme poverty, environmental degradation, pollution, corruption, and deadly wild animals. I don't think I had ever felt so alive or free in my life."

Next, she headed to Mauritius. Maggie knew very little about the country, but she rented a scooter and explored the island, making friends along the way. As she began to feel more at home, she reached out to the American embassy to find out about mapping opportunities. They connected her with the director of the Government Science Center, and Maggie was invited to give a presentation on open-source data and GIS for local students and government agencies. This presentation caught the attention of a dean at the University of Mauritius, who invited Maggie to speak to her staff about open-source data as well. Maggie spent two months in Mauritius on that first trip but had planted seeds in both South Africa and Mauritius that would continue to grow over the next few years.

Maggie explores Mauritius in 2013 on a scooter.

Maggie teaches Mauritian students about open data.

Maggie and her team monitor Lake Sibaya in South Africa in 2015 for the Save Sibaya campaign.

Maggie returned to the US but not for long. In January 2015, she headed back to Mauritius for a mapping workshop at an all-girls secondary school in partnership with the US Embassy.

Then she returned to South Africa, this time to lead Save Sibaya, a project to raise awareness about water loss in Lake Sibaya. As part of Save Sibaya, she taught groups of university students how to conduct field data collection to support a local wildlife census. By mapping the diversity of species in the area, the team hoped to draw badly needed attention to conservation around the lake and encourage the local government to step in.

The next few years would see Maggie finding GIS projects, mapping opportunities, and open-data initiatives in Saint Lucia, Jamaica, Mauritius, and Saint Vincent and the Grenadines. She continued working on the Save Sibaya project in South Africa; returned to the Naro San game reserve in Botswana with a group of students to map and track cheetah habitat, mapping paw prints and scat samples to better understand the presence of cheetah on the land; and led a student project to track and map the elusive klipspringer at Liwonde National Park in Malawi.

Tip!

"During my initial years in freelance, I found that half the battle is showing up because you really won't know what the road will be like until you take that next step. The courage to take that next step is one thing that gets me through the difficult moments."

"Success can take so many forms," she says. "When I am teaching, success is witnessing that light bulb moment. In my current role, success is about creating space for collaboration, amplifying the innovations of the OSM community, and helping to build the best open map of the world. On the road, success could mean not running out of petrol before we found the next station or getting the fire started for a hot meal."

In early 2018, she returned once again to Botswana to lead a study abroad program with a group from George Washington University. They mapped the Naro San village in OpenStreetMap, explored national parks on safari, and learned about the impact of tourism and conservation efforts in the country. She remembers, 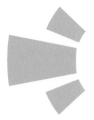 "While I was a bit worried about a student falling prey to a hyena, it was incredible to watch these American students experience that part of the world for the first time. There's nothing like watching someone's face when they encounter their first elephant in the wild."

Finding and using inner resources

Despite all these amazing experiences and loving and believing in what she was doing, there were still many challenging days on the road. "All kinds of things can go wrong," she says. "The Mozambique border guard won't let you in, the Land Rover is stuck in the sand, you don't have work for three months, the baboons get into the food supply ... but isn't that life? In the end, these experiences made me more flexible, resourceful, and appreciative." One thing she took away from urban planning is the need to get a bird's-eye view of every situation. Her ability to take a step back and broaden her perspective has helped her see the big picture and come up with alternative solutions.

Maggie continued to freelance and travel until 2019 when she unpacked her suitcase and became the executive director of

OpenStreetMap US, a nonprofit that supports the OSM project in the United States through education, advocacy, and building an active community. Now she is working to develop a strategic plan that will support the sustainability of the organization over the next few years. This position also supports Maggie's passion for open data. **"Without data, there are no maps,"** she says. **"Without data, we can't make informed decisions about our communities. Open data creates opportunities, fuels innovation, and encourages participation and diversity.**

"I feel incredibly grateful to be in the role of supporting Open-StreetMap US as it transitions from an all-volunteer organization to a staff-driven nonprofit. Although there is not a rigid framework for how the organization should evolve, the potential is so exciting because, at its core, there is not only the OpenStreetMap project but also an amazing community. I remember feeling the same way seven years ago when I started my company, Boomerang Geospatial. There are so many directions you can take, because just like GIS, OpenStreetMap intersects everything."

HANAN DARWISHE

Reaching for the stars with her feet on the ground

Position

Head of Topography Department, civil engineering
Al-Baath University in Homs, Syria

Education

PhD and master's, geomatics
Lille 1 University, France

Bachelor's in civil engineering
Al-Baath University

GROWING UP IN SYRIA, watching "the stars and the sky and dreaming of distant places," Dr. Hanan Darwishe says she was always drawn to subjects related to the earth. For her, geography lessons were like "a trip to parts of the earth on land, sea, and air. From my small seat in my school, I traveled the world and learned its secrets."

As a child, Hanan enjoyed school so much that she hoped the summer holidays would pass quickly so she could return to her classes: "For me, school was, in addition to being a place to receive science and knowledge, a place for entertainment, recreation, friendship." Hanan says she was fortunate to grow up in a family "that cares about science and encourages higher academic achievement," which motivated her to pursue her education.

"The cause I'm passionate about is learning," she says. **"So many doors can be opened and horizons widened by a passion for learning. It is reported that Albert Einstein said, 'Once you stop learning, you start dying,' and that quote has always remained with me."**

At first, Hanan wanted to be a teacher, like the inspiring teachers she'd had at school. Then she wanted to be an engineer. "Years later, I wanted to be a scientist," she says. "After a lot of thinking, I chose civil engineering and geomatics as my discipline, and I am happy with it."

Hanan studied civil engineering at Al-Baath University in Homs, Syria, where she worked so diligently that she was

Hanan, living in a country with ongoing conflict, uses GIS and remote sensing to develop GIS applications to help cities deal with damaged buildings.

ranked first in her major, surveying. "This was a victory," Hanan says, "a great achievement and a pride for me, my family, and my mother in particular, who had always stood by me and encouraged me. There were some difficulties that accompanied me during my university life, the most important of which were the difficult financial conditions that my family went through."

Today, Hanan is a faculty member of civil engineering at the same institution, where she conducts research in GIS and remote sensing, "especially in the area of developing stand-alone GIS applications to develop cities and to improve services and infrastructure," she says. Another research interest is image processing using artificial intelligence, specifically deep learning by artificial neural networks (ANNs) to classify remote sensing images, create land-cover/land-use images, or to extract a specific class, such as urban areas.

Pursuing her dream

After Hanan obtained her degree in civil engineering, she was awarded a scholarship to pursue a master's and a doctorate in GIS at Lille 1 University, France. On a personal level, the biggest challenge she faced was leaving her family and her country in 2006 to pursue these degrees. "It was difficult for me to live, for the first time, far from my mother and my family," she says. "I suffered in the beginning, especially when I left my child for three months until I found a home and registered him in a school. It was a great sacrifice for me in exchange for achieving my dream."

At Lille University, Hanan found another supportive teacher and mentor: her professor and PhD supervisor, Dr. Barbara Louche, who was "an example to follow from a scientific and humanitarian standpoint," guiding and encouraging Hanan through her studies. "She always advised me to work hard and be patient to achieve my dream. She was a sister and friend before she was a supervisor."

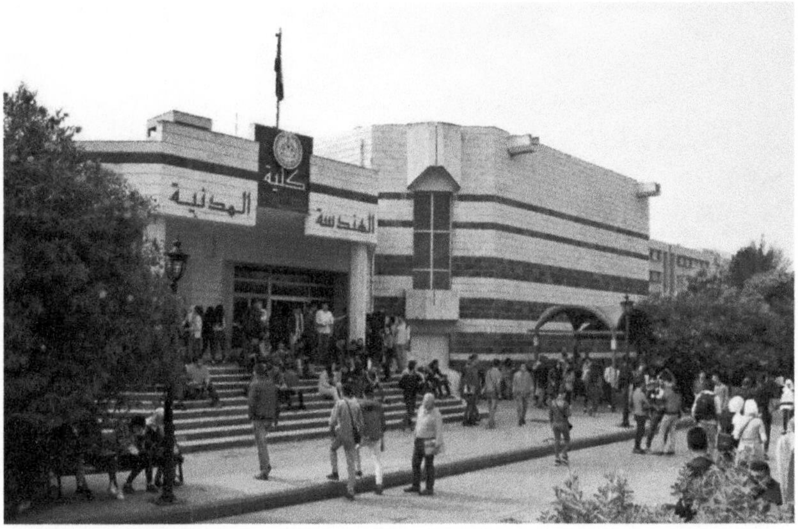

The faculty of civil engineering at Al-Baath University in Homs, Syria.

During her PhD work, Hanan participated in several conferences, the most rewarding of which was the 2010 American Water Resources Association conference in Orlando, Florida. Hanan describes it as "a wonderful scientific and cultural experience" not only for its academic nature, but also for "getting to know people who share the same passion and visiting the American continent for the first time." She also attended most of the annual Esri France conferences at the Palace of Versailles, where she enjoyed "exchanging experiences and getting acquainted with the latest Science of Where."

Fun fact!

Fun things on her desk: "A small animal-shaped toy, a rabbit toy, and a coffee mug that my sons gave me."

Although Hanan's original language is Arabic, she is fluent in French, which enabled her to teach GIS to master's students at the University of Lille from 2010 to 2011. "It was a wonderful and enriching experience," Hanan says, and it allowed her to work closely with graduate students in a field she is passionate about.

Coming home

In 2012, Hanan returned to her country, Syria, where she worked as a teacher and researcher in the faculty of civil engineering (Surveying Department) and where she began using her GIS

Damaged buildings in Homs, Syria.

expertise to solve a wide range of spatial problems. For her, the possibilities of GIS are endless, as "a science dedicated to creating the world virtually within the computer through a set of data that [represents] human life on earth, human experiences, and the relationship between man and Earth and all phenomena."

Living in Syria during an ongoing violent conflict is a challenge, but Hanan stays focused on her work because she sees it as a means of improving people's lives. She says that most of her research has "focused on making use of geomatic technologies in developing cities, especially in the period when my country, Syria, was suffering from damage and destruction to buildings and infrastructure as a result of the effects of war." She set out to "employ remote sensing and GIS in order to calculate the damage and destruction in buildings, and to prepare maps showing the percentage of damage to guide decision-makers towards taking a restoration or demolition decision," she says. She has also been responsible for training engineers in various agencies to use GIS in their specializations, such as the structural evaluation of war-affected buildings from the country's civil war.

In addition to supervising graduate students and serving as a consultant, she has worked with colleagues to develop stand-alone GIS applications to expedite and automate administrative tasks for municipalities, including the RE_3D GIS (Real Estate 3D GIS) app. The real estate evaluation process is complex and depends on

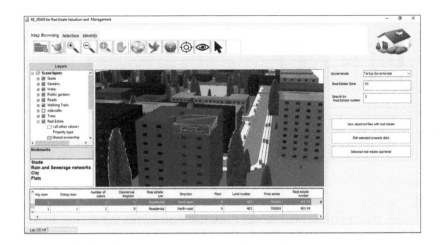

Hanan developed a desktop app for working with real estate in 3D.

many criteria that require spatial and descriptive data and smart technology capable of managing it, Hanan points out—hence, its reliance on GIS.

Taking the stairs

Hanan advises women to enter STEM fields, "as **women have a great ability to accomplish several tasks simultaneously and [use] network thinking, which suits the field of technology and mathematics with multiple ramifications.**" In her own work, she is motivated by the desire to improve the standard of life for all human beings, by putting creative ideas into practice: "We have the opportunity to shape the way humans interact with the world around them," she says. **"GIS also improves our understanding of the social and temporal factors that contribute to some of the world's most difficult problems."** But she admits that her husband, Dr. Eng. Fadi Chaaban, who is also a GIS specialist but at another university (Tishreen University in Lattakia, Syria), provides another incentive—"sometimes we try to compete with each other."

Her greatest strength, she believes, is patience: "I am meticulous and consistent. And I overcame my weaknesses with patience, hard work, and especially working as a harmonious team." Accordingly, she has another favorite motto: "There is no elevator to success—you have to take the stairs."

Fun fact!

Personal motto: "Keep your eyes on the stars and your feet on the ground."

ELENA FIELD

Charting the unknown in the Antarctic

Position

GIS and web mapping specialist
Mapping and Geographic
Information Centre at British
Antarctic Survey

Education

**MSc in geoinformation
technology and cartography**
University of Glasgow, Scotland

BSc (Hons) in geology
University of St Andrews,
Scotland

MOST PEOPLE will never visit the Antarctic, yet Elena Field has been deployed there three times. A GIS and web mapping specialist for the British Antarctic Survey (BAS), it's Elena's job to provide geospatial support and expertise to BAS operations by making maps for scientists in the field so they can travel safely. "My favorite part of the job has to be getting to visit the Antarctic," she says. "Not only was it a fantastic experience and absolute privilege, but it has helped me truly understand the challenges faced in working in such remote areas."

Her position at BAS is new to the organization—although there had been a mapping team since 1990, there hadn't been a role specifically for operations support. Because science seasons (from October to around April, when it's summer in the Southern Hemisphere and conditions are calmer) have become more complex and teams travel farther and farther afield, there was an increasing need for Elena's role. "As it was a new position, I have been able to develop it over the years, and I am still finding new ways to integrate geospatial technology into how BAS supports UK science and operations in the polar regions," Elena says.

One downside, though, is the fast pace. "Due to the constantly changing science seasons, the demands of the job and the kind of support requested vary significantly throughout the year. Sometimes it all changes all at

once—leading me and others in the geospatial team to drop everything and work on something completely different with a short turnaround time. This can be a bit of a challenge for a perfectionist," Elena says.

Working hard

In addition to her work with BAS, Elena also provides geospatial support for the UK Foreign and Commonwealth Development Office Polar Regions Department and the UK Antarctic Place-Names Committee. She also manages the South Georgia GIS, a collection of topographic, management, and scientific data about South Georgia, hosted by BAS on behalf of the Government of South Georgia and South Sandwich Islands. According to Elena, "As long as I'm learning, I'm happy."

Elena's role at BAS has allowed her to provide support on many projects, including the International Thwaites Glacier Collaboration, a five-year partnership between the US and the UK, where Elena's contributions have proved critical for field planning and have fostered constructive ways of working between BAS and the US Antarctic Program (USAP) logistics and science teams. "My primary area of research/application is how we can use GIS and

Fun facts!

Favorite trip: "My brother and I took a road trip around Iceland in 2019—it's a geologist's dream, and it was brilliant to get to experience this place. I can't wait to go back and explore some more."

On her bucket list: Traveling to the Arctic—"particularly Svalbard and Greenland. After studying and researching them for years, I'm desperate to go!"

Elena in the Antarctic in 2017 using a Ski-Doo to get around the local travel area off station.

geospatial technologies to improve the way we work, reduce impact on the environment, and complete remote fieldwork successfully and, above all, safely," she says.

Merging passions

As a child, Elena had no idea what she wanted to do, jumping from archaeologist to writer to everything in between. Although she was interested in maths and sciences, she also enjoyed being creative. **"I was always interested in the world around me, and as soon as I started studying geography in school, I was hooked,"** she says. Luckily, her career has allowed for both outlets. "I'm motivated by challenges and solving problems. **This is why I think geospatial science and technology is the perfect fit for me—there are so many options and solutions available, and it allows for a huge amount of creativity,"** she says.

Tip!

"The best advice I've been given when starting out was 'Keep at it, and you'll figure it out.' It was an offhand comment by a friend but it stuck with me."

Fun fact!

"I draw and sketch, mainly in charcoal–I mainly draw landscapes but have branched out into still-life and figure drawing more recently. I took my sketchbook to the Antarctic and have spent many hours out in the cold, drawing away."

According to Elena, growing up in northern Italy and seeing the Alps every day also fostered her interest in geography and geology. "My dad has worked at the European Commission Joint Research Centre for many years, researching and testing the efficacy of renewable energies (particularly solar)," she says. "Growing up having conversations about the environment and what we can do to help improve it has really shaped my interests."

Elena found GIS while working on her BSc in geology but preferred getting outside and learning—especially if it involved sketching landscapes. **"It wasn't until we started making geological maps ourselves that I started seriously considering cartography as a career. I just loved how we could distill weeks of field mapping and millions of years of Earth history onto one sheet of paper,"** she explains.

Making things happen

University in Scotland was a big change for Elena but a great opportunity, too. The biggest challenge—and opportunity— came during her final-year project. Her interest was in using remote sensing for geological mapping, but her supervisors were unfamiliar with the methods and her university didn't have the right software. Largely self-taught, Elena collaborated with the

Elena in northern Italy with her dog.

Elena in Scotland during her schooling.

University of Leicester and British Geological Survey, working with other researchers to develop methods for automated geological mapping based on the spectral characteristics of rock outcrops and slope analyses obtained from aerial and satellite imagery. **"Off the back of this research, I went and presented at my first conference, published my first scientific paper, and was highly commended in the Global Undergraduate Awards—so the investment really paid off,"** she says.

Elena is also passionate about promoting diversity and equality in the sciences. **"Studying and now working in a STEM field, I'm very familiar with being one of the few women in the room,"** she says. "I remember how influential it was for me to have role models when I was younger, which led me to become a STEM ambassador during my time at university." She also joined GeoBus, an educational outreach project developed by the Department of Earth and Environmental Sciences at the University of St Andrews, when it launched in 2012, and traveled across Scotland and England teaching geology and geography in primary and secondary schools until she graduated.

"If you are interested in something, don't get discouraged if it's not 'popular' or not something others are interested in–if you're interested in it, keep learning more. GIS is a fantastic tool that can be used in many different fields or applications so there is something for everyone."

During her master's, Elena discovered what would lead to her career. "It was there that I started looking into the polar regions—my thesis explored how we could use remote sensing to assess changes in permafrost coverage over time. This was my first real introduction to the challenges faced when mapping the polar regions—I was soon hooked," she says.

For as much success and accomplishment as she's had, especially so early in her career, Elena says she's had to work a lot on combating impostor syndrome. "It can sneak up on you, and it's hard to shake. This is something I'm constantly working on but having a supportive network of friends and family really helps. Not only can they give you great advice, but they can tell you to snap out of it when you need to," she says. As for what success means to her? "Honestly, I'm not sure I know the answer to this question yet," she says. "Being happy in what you are working on and knowing that you are contributing to something good and bigger than yourself," she ventures.

GABI FLEURY

Forging a path to coexistence with wildlife

Position

Conservation partnerships officer
Rainforest Trust

Education

MSc in conservation biology
University of Cape Town, South Africa

BS in geographic sciences
James Madison University, Harrisonburg, Virginia

"*I* wouldn't be doing what I'm currently doing if it wasn't for sheer willpower," says wildlife conservationist Gabi Fleury. **"As a Black scientist in a primarily White field, and as a scientist who grew up relatively lower income, I've been aware my entire career that I would have to work about 10 times harder than many others to succeed in this field."** That determination has driven Gabi throughout life—through pediatric bone cancer at age seven to a Fulbright research project in Botswana in 2021. Currently working for the Rainforest Trust—a small NGO outside Washington, DC—Gabi conducts their research in human-wildlife conflict mitigation, which involves preventing carnivores from eating farmers' livestock through the testing of different nonlethal tools and strategies.

Deeply passionate about the natural world, Gabi says working on-site and with communities is their favorite part of the job. "I usually have a strong human dimensions aspect to my work, looking at the political, cultural factors that influence farmer tolerance to large carnivores," Gabi says. So far, they've worked on or collaborated with projects based in Namibia, South Africa, Botswana, Mozambique, and Kenya. "I feel that I have a lot to learn from the individuals who share the landscape with these large carnivores, and I love exchanging ideas and collaboratively coming up with solutions together," they say.

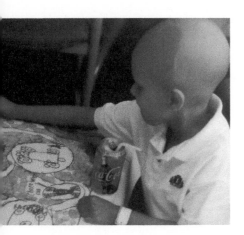

Gabi as a young animal fan at Boston Children's Hospital.

Finding their passion

The journey to becoming a conservationist hasn't been easy, but it's the dream job—aside from "pirate"—that Gabi's aspired to since childhood growing up in Boston. They even started a "Gabi goes to Africa" fund as a child because they wanted to experience more of their Angolan background and culture. "I have no clear memory of the moment when I decided to become a conservationist, but I expect that it grew out of feeling a strong connection to my heritage and loving wildlife, particularly big cats and African wild dogs, and wanting to protect them," Gabi says.

At seven, though, a pediatric bone cancer diagnosis changed everything. Gabi spent several years going through 21 rounds of chemotherapy. No longer able to run or play, and going through extensive physical therapy to learn how to walk again, Gabi's interest in wildlife blossomed through time spent reading in the hospital. They consumed massive wildlife encyclopedias and memorized Latin names and animal facts.

"I knew growing up that if I wanted to do conservation, it was something that I would have to figure out how to do myself," they say. "I never had the opportunity to meet a field conservationist as a child, and I knew that education and working abroad was expensive, so I've always been very driven to make that career happen for myself. If I was going to work on Southern Africa conservation issues, it was going to be under my own steam."

Overcoming obstacles

Gabi finally made it to South Africa after winning an Ambassadorial Scholarship from Rotary International for an accelerated master's degree program at University of Cape Town in conservation biology. "My master's experience taught me the patience, flexibility, and grit necessary to succeed in scientific research, sharpened my spatial and statistical analysis skills, and fully engaged me in the

Gabi during field study at the University of Cape Town in South Africa.

South African conservation community, which I have found essential throughout my research career," Gabi says.

From there, Gabi started as a field tech, trekking in the Northern Cape of South Africa over crumbling rocks to collect data on livestock loss from leopards and caracals. This was not an easy feat—as a result of their childhood cancer, Gabi has a reconstructed lower left leg. It took years of physical therapy and hard work to even walk. Now, though, they can do hours-long hikes in various terrains with the use of only a small ankle brace.

But physical obstacles aren't the only thing Gabi has had to overcome to make it in their dream job. Conservation, especially conservation that focuses on charismatic megafauna (big cats, whales, and so on), is competitive. And because of the skill and education requirements for jobs in conservation, Gabi says the discipline favors those who can afford to pay for internships and field courses, which impacts diversity. "There's been many late nights of staying up reading scientific literature on carnivore conservation and networking with the authors; creating fund-raising campaigns for projects like my human-wildlife conflict video game, Operation

Tip!

"Best advice I was given starting out was 'forge your own path.' Conservation isn't a structured, straight-line career; you can get into it in many ways. This is extremely exciting, but it also can be really challenging, because you have to be flexible, innovative, and always on the lookout for the next opportunity. It's not for everyone."

Gabi during an internship at Big Life Foundation in Kenya.

Ferdinand; self-teaching myself R, quantum GIS, video-editing, database creation, Photoshop, and Python to build my skill sets; and grant writing or taking odd jobs to get myself to my next conference or field site," they say.

Making things happen

When Gabi's internship ended, they spent their time networking (often through emails) with conservation organizations, working for a stipend or for free on-site in Africa for several months, and splitting time in the US, working hard at retail jobs to fund their next plane ticket. As Gabi says, **"I never give up once I set a goal in mind, even when the going gets hard or I have to adjust, innovate, or change course to make it happen."** That mindset inspired Gabi to trade spatial, statistical, and grant-writing skill sets to convince NGOs that they'd be an asset to the team and thus built their on-site experience that way. It worked, resulting in an

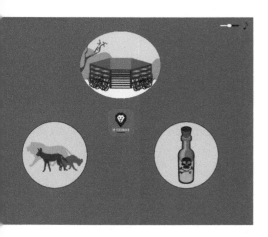

The interface of Operation Ferdinand, the video game Gabi helped create.

internship with Big Life Foundation in Amboseli, Kenya, working on its Predator Compensation Fund and with the same Maasai community that Gabi worked with for their undergraduate thesis. They also did grant-writing workshops and taught R programming in exchange for on-site experience with another small grassroots conservation organization based in Salama, Kenya.

On the side, Gabi collaborated with a software engineer colleague to design and create Operation Ferdinand, an environmental education video game comprising three mini-games composed of pictures only and no text to allow for learning regardless of language barriers or literary levels. "These mini-games taught carnivore identification based on tracks and scat, nonlethal ways to prevent livestock loss to predators, and the ecosystem and human health effects of poisoning carcasses to kill carnivores," they say. "Operation Ferdinand was collaboratively developed and play-tested with the local NGO, Niassa Carnivore Project, in northern Mozambique as part of their environmental education program with children and young adults from the surrounding villages in the Niassa Reserve."

Gabi is also passionate about increasing representation in STEM. "Positive representation of Black and LGBTQIA+ scholars is so crucial and is an additional driving factor into why I have devoted my life to a research career in the biological sciences. I hope that it may open up doors for those who come after me," Gabi says. They cofounded a science communication YouTube channel called Breaking Bio, where they interview STEM practitioners from underrepresented groups. They're also involved in Skype a Scientist, an initiative that pairs scientists with elementary school classes to speak about their research. And Gabi's an inaugural member of the organizing committee of Black Mammologists Week, an annual campaign on Twitter and Instagram, in partnership with the National Geographic Society, that was created to raise awareness of the historical and current work of Black mammologists across the diaspora.

Finding success

All the hard work and perseverance has been worth it for Gabi, who was named a 2021 *Forbes* "30 under 30" mover and shaker. They say they wouldn't trade any of it for the world. "I feel confident in saying, at 28, that I've made it," they say. "I've gotten two degrees with no debt and am hopefully on track for a third, have published in the scientific literature, built my contacts in South Africa, and have several long-term paid positions under my belt, but it was incredibly hard to get there." They sacrificed time and effort to build a career and lived abroad in difficult field conditions but finding sustainable ways for humans and wildlife to coexist is the driving factor behind everything Gabi does.

Next up for Gabi is a Fulbright research project in Botswana in collaboration with Botswanan NGOs Cheetah Conservation Botswana and Botswana Predator Conservation to test the use of scent deterrents to keep subdominant carnivores such as African wild dogs, one of Gabi's initial loves, off commercial farms.

And for young women looking to enter STEM fields, Gabi encourages: "You can do anything you set your mind to. It might be a bit of an uphill battle, as women still are less represented in STEM, but there are so many women excelling in their fields, and we need young people to continue to work to change the world with their creativity, brilliance, and drive."

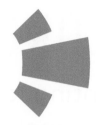

Gabi in 2020, working from home.

AFRICA FLORES-ANDERSON

Fighting for a sustainable world, from Guatemala to the Himalayas

Position

Land-cover and land-use change thematic lead and regional science coordination lead
SERVIR-Amazonia

Education

Pursuing PhD
McGill University, Montreal, Quebec, Canada

MS in earth system sciences
University of Alabama in Huntsville

BS in agronomy engineering in renewable natural resources
Universidad de San Carlos in Guatemala

As a child in Guatemala, Africa Flores grew up hearing stories about pristine tropical rain forest landscapes with clean fresh water and wildlife living close to villages. But smoke, ash, and polluted rivers replaced those natural wonders before she ever saw them in Retalhuleu, a town in southwestern Guatemala that's a major hot spot of industrial sugar cane production. This reality sparked Africa's passion for the environmental sciences so that she could learn how to sustainably manage natural resources.

Now Africa, a research scientist at the University of Alabama in Huntsville and a National Geographic Explorer, does just that—working with satellite observations and geospatial technologies to improve decision-making in SERVIR regions. SERVIR is a joint program between NASA and the US Agency for International Development (USAID) that partners with centers of excellence in eastern and southern Africa, West Africa, Hindu Kush Himalaya, and the lower Mekong and Amazonia. In her current role, she oversees SERVIR's global land-cover and land-use change portfolio and leads cross-region and cross-theme scientific collaboration on land-cover activities, knowledge sharing, and sharing of best practices. Although her work is

Africa as a child in Guatemala at age 3.

demanding and time consuming, Africa says, "I love that I'm constantly learning and that my work is challenging and exciting."

In addition to serving as coinvestigator in multiple NASA-funded projects, Africa is the principal investigator of a project supported by National Geographic and Microsoft. "I remember when my dad brought the *Nat Geo* magazines to our house, and my brothers and I would devour them," she says. "Now I am a Nat Geo Explorer using artificial intelligence to forecast harmful algal blooms in Lake Atitlán, Guatemala. It is definitely a highlight in my career."

Pursuing an education

Education and learning are two things that have been a constant for Africa. When asked as a child what she wanted to do when she grew up, her answer was, "I want to keep learning," she recalls. "I think that showed my early interest to be a scientist. But I didn't know that learning was even a career because I couldn't see any examples of this close to me or my reality."

Africa as Abanderada General (first place for the whole school) in Retalhuleu, Guatemala.

Though she had limited opportunities available while growing up, Africa was happy just to get an education—something denied or not accessible to her mother and grandmothers. "I understood at a very young age that education was a privilege and that I had to make the most of it," she says. "Education was my opportunity to one day become a professional and be financially independent, to break the cycle of financial dependence that women are usually subjected to, particularly in countries like Guatemala." Determined not to waste her opportunity, she worked hard and succeeded, despite challenges such as suffering bullying and economic hardships.

"The sacrifices my family did to keep me in school and give me an education were enough fuel for me to keep trying to do my best at school. My parents made the whole difference, and they always supported me so much," she says.

Africa's graduation in agronomy engineering in natural renewable resources from Universidad de San Carlos.

For university, Africa attended Universidad de San Carlos (USAC) in Guatemala on scholarship. "Our economic situation was so precarious that I could not afford to live in the city to go to university. I'm eternally grateful to USAC and their scholarship system. I could not have had access to quality higher education if it were not for USAC. **I would not be a professional today if public higher education in Guatemala didn't exist,**" she explains. Now Africa is pursuing her PhD at McGill University in Montreal, Quebec, Canada, while working at SERVIR.

The biggest hurdle, though, was overcoming the discrimination and machismo embedded in Guatemalan society, Africa says. The first woman to top the faculty dean's list since its creation, she faced harassment and bullying. **"I thought, if this is the price to pay to get an education, I will pay it. It doesn't compare to the price of not getting an education at all. I very well know what that price of not going to school is, and my family and I wanted something different,"** she says. "I realize that wherever I would have gone in Guatemala, I would have experienced something similar, because my sins were to be a woman and be poor. Ultimately, these experiences made me stronger and helped me develop emotional maturity." In the end, Africa would become an agronomy engineer with a focus on renewable natural resources—the closest to environmental sciences she could find in Guatemala.

"My persistence and obstinacy to keep going have brought me to where I am today," she says. **"Do not give up. Keep going."**

Starting her career

After university, Africa joined the National Protected Areas Council in Guatemala, starting as a GIS specialist before then leading the GIS unit. She went on to become a research scientist at the Water Center for the Humid Tropics of Latin America and the Caribbean, in Panama, working in SERVIR. From there, she moved to the US for a master of science in earth system sciences at the University of Alabama in Huntsville and worked with SERVIR as a graduate research assistant. After graduating, Africa continued working with SERVIR, coordinating activities in eastern and southern Africa and the Hindu Kush Himalayan region.

A major milestone in her career was leading the SERVIR global effort to build capacity in the use of synthetic aperture radar (SAR) for ecosystem applications, a collaboration with SilvaCarbon. This led to the creation of the *SAR Handbook: Comprehensive Methodologies for Forest Monitoring and Biomass Estimation*, which Africa coedited and coauthored. "It has been extremely well received since it provides practical examples on how to use SAR to map forests and their condition," Africa says. "In a little over a year from its release, the handbook and related materials have been accessed over 500,000 times from all over the world."

Advocating for women

Being named the 2020 Geospatial World's Geospatial Woman Champion of the Year, Africa says, "instills in me yet a greater sense of responsibility to create conditions for women to succeed in fields related to GIS and remote sensing." It's a cause she's passionate about and one that's personal, too. **"STEM fields including GIS are fun, interesting, challenging, and they need women's perspectives and skills.**

"The constant need to prove myself in my field has been a struggle and a frustration. To be strong in the face of a subtle, unconscious bias against women and women of color, found in science, can be exhausting. Sometimes it feels like swimming against the current. Then, you get used to it, you learn, and find strategies to cope with it. But the moment you think you have everything under control and experience bias again, it can be tiresome."

Africa says success to her means "to be fulfilled both in my personal and in my professional life. To be happy and loved."

Africa presenting at NASA Hyperwall at the American Geophysical Union about the *SAR Handbook* in 2019.

Source: Emil Cherrington.

One of the hardest choices Africa has faced was whether to continue working after having her first child. "It was the first time I questioned staying in the field. My kid needed me, and this work can be too demanding. On top of that, when you add some of the struggles that your male counterparts do not face, you really question if it is worth it," Africa says. "Other female colleagues in my field proved to me that it is possible to stay. Their words of support and their example showed me that it is possible to be both a mother and a scientist. I admire their integrity, strength, smarts, and legacy. Nancy Searby, Sylvia Wilson, and Jenny Frankel-Reed are among those women. Thank you—you have taught, and continue to teach, me by example."

Africa also hopes that her career has contributed to making geospatial technologies more accessible to communities around the globe. "My passion is to empower the people who depend on these natural resources, because it's they who will make a change," she says. "It is their land and their water. It is their future."

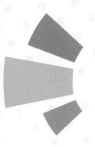

MIRIAM GONZÁLEZ

Democratizing access to geospatial data

Position

In Partnerships for UP42

Education

Diploma in geomatics
Universidad Nacional Autónoma
de México

Master's in international
business
University of Groningen, the
Netherlands, and Skema
Business School in France

Bachelor's in business
administration
Universidad Iberoamericana,
Mexico City

*G*ROWING UP IN THE RAIN FOREST in a house surrounded by mango and avocado trees, just 30 minutes from the sea, Miriam González, who goes by @Mapanauta, loved to explore nature. At school, she enjoyed science experiments. But, she says, "I also remember watching Carl Sagan speaking in Spanish (translated to Mexican Spanish) on the TV. He was the best teacher a kid can have."

As a child, Miriam believed that everything was green. She remembers the green of the jungle of Veracruz, so bright that it hurt her eyes, and her surprise to find out there were places that weren't so lush and fertile. She recalls, "When I was only seven years old, my family went to live for a short period in the Coahuila Desert. The long walks losing myself in the desert immensity showed me the first freedom sensation that I am addicted to today. **When I started traveling, one of the most wonderful feelings I had when I came back to Mexico was seeing how the Iztaccihuatl and Popocatépetl Volcanoes are like a giant couple on the horizon that welcome you home. Mexico has showed me to love nature, culture, and history and has given me so many nice memories, which are like living tattoos.**"

Both of Miriam's parents were abandoned as children, so they had to start working at a young age to support their younger siblings. Her dad managed to work and get a degree in medicine and is now a retired doctor. Her mother is a businesswoman, and as Miriam says, "At 73, she keeps reinventing herself. She could sell sand to the deserts in the [United Arab] Emirates."

Miriam was the first of two children, and at 11 years old, she started helping in the family business. Spending her afternoons working and interacting with customers fit her perfectly as an introverted child and shaped her personality today.

Inspired, perhaps, by Sagan, Miriam has a vision for her future—not a dream, she says, but "a fact: I am going to space." She doesn't know when, but she's sure she will travel there one day. She has already visited some amazing places, from the depths of the ocean (she is a scientific scuba diver) to Antarctica, which,

Miriam says Antarctica is one of her favorite places.

she says, is her favorite trip so far: "I am still dreaming about all the memories captured in my head." Also, she is learning how to sail, planning "in the next few years some good sailing trips around the globe."

In her professional life, Miriam works in Partnerships for UP42, a developer platform and marketplace for geospatial data and analytics that aims to solve Earth's challenges, and says that her goal is "to democratize access to space and geospatial data and work with revolutionary companies and organizations which are creating algorithms that can help to solve Earth's challenges."

She started her career with a background in business. After undergraduate work at Universidad Iberoamericana in Mexico City, she studied for a master's in international business from a consortium of European universities. "I spent most of the master's time in the south of France," she says, "and it was a great combination of working a lot but also enjoying the international student environment." Plus, she took a sabbatical year to study Mandarin at Beijing Language and Culture University in Beijing, China: "It was the most challenging thing to learn," she says.

Though Miriam was always interested in business, influenced by her mother, she also loved science and nature, and when it came to a career, "the only thing I knew for sure is that I wanted to feel free at work doing something that keeps me happy every day. I have been very lucky with that, and being able to travel the seven continents while working remotely in the past decade is something I enjoy about the things I do."

Miriam's introduction to geospatial technology was in her previous position, where she was responsible for bringing the first GPS navigation app in smartphone devices to the Latin American market. A few years later, she became aware of open geospatial data in OpenStreetMap and started supporting humanitarian mapping through Humanitarian OpenStreetMap Team (HOT), an international group dedicated to humanitarian action and community

Two-year-old Miriam with her dad, Victor Manuel González Montañez, when he was studying to be a doctor.

Miriam with her parents, who are a source of inspiration for her.

development through open mapping. When she investigated the state of the map in Latin America, she decided she could help create awareness about the importance of open mapping. Now, she says, "I am an advocate of open mapping and open-source geospatial tools."

In her volunteer work, Miriam is president of the board of directors of Humanitarian OpenStreetMap and a cofounder of Geochicas. "What we, as a group, have achieved in Geochicas makes me proud every day," she says.

Geochicas started as an initiative in Latin America but is now global and has more than 200 members worldwide. According to Miriam, Geochicas has three main goals:

- Increasing the number of female mappers in OpenStreetMap. Surveys indicate that 97 percent of all mappers are male. Geochicas wants to create the most accurate map by representing gender reality, Miriam says.
- Creating a network to encourage and mentor women, resulting in more women leading geospatial projects and having more female speakers at global events.
- Sharing knowledge so more women can access new career paths or improve their positions.

"We can't wait to be able to gather in person again," says Miriam. "At the beginning of geospatial conferences, we organized an ice-breaker event named 'Geochicas Take (NameOfTheCity),' and it created a wonderful welcoming feeling for all the newcomers. So many projects and institutional cooperations were born after creating this community."

Miriam says that she is passionate about "democratizing data accessibility, having diversity in the mappers who are adding data to open mapping, diversity in the creators of open-source tools, and motivating more women to be part of STEM efforts in different areas—there is so much to learn and to share." For young women

Miriam with bananas she cut in her family's backyard, where she remembers the brilliant green vegetation in the rain forest of Veracruz, Mexico.

Tip!

"Enjoy what you are doing, and don't forget to smile."

who are considering entering STEM fields or exploring GIS, she says: **"Keep being curious and never stop learning new things. Don't be afraid to make mistakes—those mistakes will become great experiences, and you will laugh about them later."**

Her greatest strength, she believes, is that she is persistent and likes to make things happen. "When I have a challenge, immediately my mind starts creating options for solutions, something like A-Z options. This is one of my biggest strengths. My friends always get surprised about the craziest options I give them to solve issues."

Miriam scuba dives opposite a manta ray at the Revillagigedo Archipelago.

Courtesy of Yuko Yoshikawa.

Also, she says, "I never stay still—that can be good but also can be tiring sometimes."

For Miriam, at this stage of her life, success is defined by autonomy and mobility: "Be doing what you love, surrounded by wonderful people, and working from wherever you are in the world with a good Wi-Fi connection. What else do you need?"

Keep up with Miriam's work by following her on Twitter at @Mapanauta.

HEALY HAMILTON

Answering life's call to help save the diversity of life

Position

Chief scientist
NatureServe

Education

PhD in integrative biology
University of California, Berkeley

MA in environmental studies
Yale University

BA in ecology, behavior, and evolution
University of California,
San Diego

*I*N HER EIGHTH-GRADE YEARBOOK, Dr. Healy Hamilton listed her future profession as "oceanographical environmental lawyer." At the time, many nations were still hunting whales with exploding harpoon tips and killing tens of thousands of dolphins caught in the nets of tuna fisheries. Healy was so outraged by these practices that she dreamed of becoming a lawyer to fight on behalf of the dolphins and whales. But she never considered a career in science. All through school, even on the gifted-student track, she wasn't encouraged to pursue science. In high school, she says, she took "the absolute minimum of science and never took it very seriously or appreciated its value."

Today, though, Healy is the chief scientist at Nature-Serve, the world's first biodiversity information network, and a world-renowned scientist in biodiversity conservation. She is also a world expert on the taxonomy, evolution, and conservation genetics of sea horses, sea dragons, and pipefish. For the past 20 years, her field of research has been applied biodiversity informatics. If you're not sure what that is, Healy says, you're not alone. Biodiversity informatics is the creation, integration, analysis, and understanding of information regarding biological diversity. Applied biodiversity informatics is focused on information that has practical meaning to the management and conservation of the diversity of life.

Healy encourages everyone, especially young women, to follow what they're interested in, even if they don't know how it will play out in a career.

Healy's journey to her current position was, as she describes it, "serpentine"—winding and indirect. She grew up in central Marin County, California, with access to the abundant wildlife and open spaces of places such as Mount Tamalpais State Park, the Golden Gate National Recreation Area, and Point Reyes National Seashore. And, Healy says, her mother was an environmentalist before the term was even defined. "As a family, we spent Mother's Day taking care of our ultimate mother—our planet—with beach cleanups and invasive species removal," Healy says. Although she didn't realize it at the time, Healy grew up with a combination of family values and experience of the natural world that shaped the environmental scientist she would become.

Healy admits to being "a rebellious teenager," without much interest in academics, so she didn't go straight to college after high school. Instead, she worked and traveled for two years before finally returning to begin her college education at the University of California, San Diego. These gap year experiences gave her a "pretty deep sense of independence and confidence," she says. "It was not the first time I had lived on my own, and I was in college by choice."

A life-changing lecture

Still thinking she might be an environmental lawyer, Healy entered college as a political science major. But in the first semester of her sophomore year, she enrolled in a science course for nonmajors, Ecology and Man, to fulfill a requirement. During the first lecture, the professor discussed major global environmental issues and the urgency of dealing with them. For Healy, that single class was life changing. **"After that first lecture," she says, "I literally sprinted down to the registrar's office to change my major to ecology—having no idea what I was getting myself into. But I never looked back, eventually going on to obtain a master's and then a PhD."**

Healy finished her doctoral degree when she was in her mid-30s but still wasn't sure what direction her career would take. She'd had few female mentors throughout her higher education, and much of the advice she received—such as the importance of becoming highly specialized in one field—was contrary to the path she ended up taking. What she really wanted was a job where she could do, in her words, "biodiversity research, conservation, and education and outreach—all rolled into one position."

An ideal first job

One day she went to visit a former student at the lab where she was pursuing her master's degree, at a major natural history museum. As Healy walked up the museum steps, she realized that this was the type of institution where she could have a job that involved all the elements she was hoping for. Taking the initiative, she wrote a proposal to start an applied biodiversity research center at the museum and persisted until she gained the curators' interest and support. Eventually, they agreed to let her launch the center if she could raise the funding—and she did. "Navigating by instinct," Healy started her professional career by creating her ideal job.

Healy is a world expert on the taxonomy, evolution, and conservation genetics of sea horses, sea dragons, and pipefish. Here, she is holding a pregnant male *Hippocampus abdominalis*, or pot-bellied seahorse during one of many field expeditions to Australia, the center of diversity for this unusual and charismatic fish group.

"One of the most wonderful things about being a biodiversity scientist is going to remote places to explore and understand nature," Healy says. Here, she is on a small uninhabited atoll in the Tuamotus Archipelago, a vast chain of scattered islands in the South Pacific that is part of French Polynesia.

She spent the next decade at the California Academy of Sciences natural history museum, building a diverse lab that integrated geospatial and biodiversity science for conservation applications, while also training a cadre of Latin American and female scientists and engaging with the museum's many public audiences. **"It was," she says, "the most fulfilling first job I could ever have imagined."**

Heading east

Moving to the Washington, DC, area to take on the position of NatureServe chief scientist was, Healy says, "a pretty huge personal sacrifice." Her family, her community of close friends, and the instant access to open spaces that her life in Northern California had offered was a lot to leave behind. But the professional challenge, the leadership opportunity, and the potential impact were all compelling reasons for her to make the leap to the East Coast. Healy says that from a young age, she has had confidence in her instincts and the courage to follow them, even when they lead her in unexpected directions.

As she'd hoped, her current position has provided an "amazing opportunity" to grow her leadership skills and expand her impact. To protect threatened biodiversity, NatureServe works with the NatureServe network, a collection of 100-plus organizations and more than 1,000 conservation scientists across the Western Hemisphere. Each member program is staffed with zoologists, botanists, ecologists, spatial analysts, and data scientists who collect high-quality information about biodiversity, with an emphasis on species and ecosystems that are imperiled. "I lead an outstanding team of dedicated biodiversity scientists," Healy says, "and our work is used by all major federal resource management agencies, the private sector such as forestry and infrastructure, other NGOs, academics, educators, and the general public."

These days, Healy says, she is focused on "how to leverage our unique high-quality biodiversity observation data by integrating it with modern tools of ecological modeling and cloud computing." The field of ecological modeling has advanced rapidly, she says, which only increases the value of the type of precise, reliable field observation data that the NatureServe network has been creating for nearly five decades.

Alarming trends

Healy admits that her work at NatureServe is rewarding but also extremely challenging. The most challenging aspect of her job, she says, is "how understaffed and underresourced we are relative to the challenge and urgency of our mission." One of her greatest frustrations is what she calls a lack of "basic ecological literacy" at all levels of society. Most people, she believes, are not aware of, or

In 2006, Healy led an expedition of marine scientists to investigate the underwater diversity of Palmyra Atoll National Wildlife Refuge, a small island in the Central Pacific that is managed as a public-private partnership between the US Fish and Wildlife Service and The Nature Conservancy.

Left: Healy decries the public's seeming lack of "basic ecological literacy."

Right: Healy, left, and Regan Smyth (also featured in this book) present the Map of Biodiversity Importance, developed by NatureServe, at the 2020 Esri Federal GIS Conference.

not paying attention to, the alarming trends of decline in biodiversity and ecosystems. And yet **"this fundamental, critical issue of the degradation of our planet's ecological health has everything to do with current and future human well-being."** How something so crucially important can be so ignored is, Healy says, "a huge frustration."

In spite of the challenges, Healy remains highly motivated and committed to the cause of biodiversity conservation, driven, she says, by the sense of meaning she finds in her work:

> **In my value system, sustaining the diversity of life and protecting an ecologically intact natural world are the most important causes there are. My job allows me to actively contribute to these causes in a tangible way. This is the difference between a job that is a professional career and a job that is a way of life.**

In addition to that profound sense of meaning, Healy says she loves "the intellectual challenge of designing the science that fuels informed decisions." And, she says, "I am also reminded every day what an extraordinary group of colleagues and partners I have the pleasure of interacting with—this is a field that attracts special people that are passionately committed to conserving life on earth."

Today, Healy believes that we are at an extraordinary moment in human history. Even before the COVID-19 pandemic, we were at "a moment of extreme awareness regarding the consequences of the trajectory we are on—the negative consequences of exceeding Earth's planetary boundaries." Many efforts to change that trajectory have fallen flat to a large degree. But now, the pandemic has shown us that, in fact, "dramatic societal change can happen almost instantaneously on a global scale."

To be part of such change, she says, young women should be inspired and encouraged to consider a career in data science. Why data science and GIS? For Healy, it's a no-brainer: **"There are few more important careers than to seize the data and tools of the moment to help create planetary awareness and build ecological literacy.** The abundance of data sources will only increase and diversify in the future. The tools of spatial science can bring order and meaning to this data for the scientist and nonscientist alike. **GIS can bridge data science, art, and storytelling—a combination that can uniquely change hearts and minds. What other career choice can say that?"**

Tip!

"Success, to me, is a daily feeling that you are living true to your values, that you are meaningfully contributing to the world you want to create."

KATHARINE HAYHOE

Spreading the word on climate change—and action

O NE OF DR. KATHARINE HAYHOE'S professed greatest strengths is that she is a generalist. She advises graduate students in disciplines from psychology and public relations to engineering and political science. She collaborates with colleagues who publish work in journals of education, transportation, and agriculture. She teaches classes that are open to students across the university, including those studying architecture, law, and medicine. A generalist is exactly what someone who has devoted their career to studying climate change needs to be, she says, as it affects nearly every aspect of life on this planet, and we need nearly every skill set we have to tackle it.

Katharine is an endowed professor in the Department of Political Science at Texas Tech University, and in 2021 she was also named chief scientist at The Nature Conservancy, a global environmental organization. Katharine says her research focuses on analyzing long-term observations to see what's already changed, and then developing high-resolution climate projections to show how climate will continue to change in the future depending on the choices we make. She also works with engineers, planners, and other authorities to figure out how to use this information to make our food, water, and infrastructure systems more resilient to future climate risks.

Position

Climate scientist
Texas Tech University

Chief scientist
The Nature Conservancy

Endowed professor of public policy and public law
Department of Political Science at Texas Tech University

Education

PhD and MS in atmospheric science
University of Illinois at Urbana-Champaign

BSc in physics and astronomy
University of Toronto, Canada

Katharine is finishing off a huge new dataset of super-localized climate projections that any city across North and Central America can use to understand how climate change will affect their temperature and rainfall in coming decades. And on top of that, Katharine says, "I've recently finished a book called *Saving Us: A Climate Scientist's Case for Hope and Healing in a Divided World* [forthcoming in September] that talks about why climate change is so contentious and how we can move past the political polarization. And finally, we're currently producing the final season of episodes for my PBS digital series on YouTube, *Global Weirding*."

Advocating for change

"I see my work as similar to that of a cardiologist or a pulmonologist who, after doing a scan, might tell you that your arteries are already partially blocked or your lungs have some spots on them, but if you improve your diet and lifestyle, you could still lead a long and healthy life," says Katharine. "If you continue your current lifestyle, though, they warn that you could experience some serious and even dangerous consequences. Your future is in your hands."

Katharine decided to spend her life studying climate change and advocating for climate action because, she says, "Climate change is not fair. Its impacts disproportionately affect the poorest and most vulnerable people on the planet, the very ones who have done the least to cause the problem in the first place."

Climate change is a threat multiplier, she says. It exacerbates the preexisting and fundamental stressors of poverty, hunger, and systematic lack of access to basic health care, education, and even clean water, Katharine says. It amplifies gender inequality, racism, the marginalization of indigenous peoples, and more, she adds. Katharine is passionate about climate change because its impacts are so profoundly unjust, and because there are so many solutions and organizations that can address these immediate needs

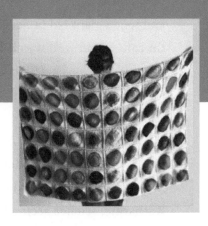

This blanket or throw is one of Katharine's latest knitting projects.

and inequalities: providing clean energy to communities without power, improving agricultural production, cleaning up air pollution, and empowering people across the world, through everything from microloans to education. **"Climate solutions won't just fix climate change, they'll fix a lot of other problems along the way,"** she stresses.

A love of science starts early

Katharine's father was a science teacher, so she grew up thinking that science was the coolest thing you could possibly learn about. She loved reading as a child, and she originally wanted to be a librarian so she could read all the latest books as soon as they came out, for free. Katharine didn't have a television, as both of her parents believed television stunted children's brain development. She remembers, "But every Friday, my mom would go to the library and rent a giant old projector and films—real films, in huge round tin cans—and we'd watch Errol Flynn, Julie Andrews, and documentaries about Jane Goodall and her chimpanzees. Jane's passion for science—and her example of being a young woman doing it— still inspires me today."

Katharine with Jane Goodall, a childhood inspiration.

Every summer as a child, Katharine had a science-related project: finding native wildflowers in the forest, learning to identify birds and their calls, or becoming fluent in binary numbers. To Katharine, science seemed like a natural way to understand the world. "By the time it got challenging and dry, like during my statistical mechanics and electricity and magnetism courses in university, it was already years too late: I was hooked," she says.

When Katharine was nine years old, her family moved to the city of Cali, Colombia. Living there altered her perspective on the world. Although most of her friends at the school she attended were relatively well off, many of the friends she made at church and in the community were living well below the poverty line. "For

Fishing at the cottage where her family spent summers during her childhood.

Katharine in Colombia, where her family lived for many years.

them, a car was an unusual luxury, and if you had one, you'd give rides to everyone you knew whenever you were driving somewhere," she says. "A home was something you built yourself over years, with only the bricks you could afford to buy with each paycheck. And when disaster hit, you had little or no recourse for recovery. They taught me that you don't need much of what we in rich countries consider necessary to be happy, but at the same time, I could clearly see that the less you have, the more vulnerable you are."

An unconventional career

Katharine began her later academic career pursuing a bachelor of science in physics and astronomy from the University of Toronto. Her first published papers were in the field of observational astronomy, on variable stars and galaxy clustering around quasars. As she was finishing her degree, she took a class in climate science with Danny Harvey, who had previously been a post-doctoral scholar at the National Center for Atmospheric Research

Fun fact!

Favorite trip: "Five years ago, I was invited to go to Churchill, a small town in northern Manitoba, on Hudson's Bay, where the polar bears congregate in summer. Polar Bears International asked if I wanted to come with them to observe the bears out on the tundra as they prepared for their annual migration out onto the ice in late October. I wanted to – who wouldn't? – but I study how climate change affects the places where we live, in ways that matter to people, so I wasn't sure that this was the best use of my time. But when I said this to Steve Amstrup, their lead scientist, he replied with some words of wisdom I've never forgotten:

'We study the bears,' he said, 'because after them, we're next.'

"So I went to Churchill, and I kept that in mind the whole time I was there, and both his words and that experience have forever changed the way I think about climate impacts and our shared future."

Spending time in the great outdoors boosts Katharine's well-being.

with Steve Schneider. She says, **"That class completely shocked me and ended up changing my life. I didn't realize climate science was based on the exact same basic physics—thermodynamics, nonlinear fluid dynamics, and radiative transfer—I'd been learning in astrophysics. And I definitely didn't realize that climate change wasn't just an environmental issue."** She switched gears and headed to the University of Illinois at Urbana-Champaign to work on a master of science in atmospheric science.

In addition to the challenges of switching fields, she vividly remembers when, thanks to some challenging (and not so interesting) math and physics classes, her GPA ended up being just one tenth of a point below the cutoff threshold required to renew the scholarship paying for her tuition. She pleaded with all her professors to see if they'd increase her grade by enough for her to make the cut, but it didn't happen.

She remembers:

I didn't know what to do; I was paying for my tuition myself, and the money I made from working my part-time summer job wouldn't cover it. I was in despair … and then I received an invitation to meet with the chair of our department. I had never met someone as imposing as the chair of our entire department before, and I had no idea why he might be calling me into his office.

He asked me how I was doing. Nearly in tears, I confessed that I had just missed the cut for my scholarship. He asked me to consider applying for one of the National Science and Engineering Research Council's undergraduate research scholarships. The monthly stipend would cover tuition and provided valuable research experience for graduate school as well.

He handed me an application, and I applied that same week. Within days, my application was accepted by an astronomy professor who wanted me to monitor variable star brightness from the rooftop telescope. The next year, I was analyzing CCD images from the Canada-France-Hawaii telescope, and the summer after that, I was testing prototypes for a new satellite instrument to measure methane in the atmosphere. These experiences shaped my life and made me grateful that I'd lost the scholarship because I'd gained so much more.

Seeing the light

Katharine's advice for any woman interested in the STEM fields: "Data is what illuminates our world. Whatever your interests, whether in the arts, humanities, social sciences, physical sciences, engineering, or beyond, understanding and being able to work with data will enhance what you know about this world we live in and how you as an individual can navigate it."

She loves seeing climate information used to make a difference in people's lives, from farmers and agricultural producers in rural areas to urban planners concerned about social inequalities and resilience in some of the world's biggest cities. Her least favorite part of being a climate scientist, she says, is the abuse she receives from people who don't want to acknowledge that the climate is changing, that humans are responsible, that the impacts are serious, and that we need to act now. Every week, she says, she hears criticism from people and disbelievers via mail, email, social media, and phone calls.

"In a way, though, these attacks are sort of perversely encouraging," says Katharine. "Would those opposed to climate action be so vocal and so hostile if they didn't think I was making a difference? Hopefully not!"

> **Tip!**
>
> "While it's important to have people you respect and trust give you feedback at key points in your career, when it all comes down to it, you have to make the decisions that feel right for you, not the ones that necessarily look best on paper. You're the one who has to live with them."

A measure of success

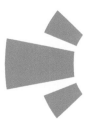

Katharine speaks at a variety of forums around the world.

Katharine has received many honors throughout her career and has had the opportunity to speak at a wide variety of forums, from the World Economic Forum to international science festivals and from top universities around the world to the annual Esri user conference.

"I don't measure my success based on these recognitions or invitations, though, because they depend on other people's opinions of me and of what I do," she says. "And while I can influence them to some degree, I'm not ultimately responsible for what others think of me.

"To me, success means that I've accomplished a goal that I've set for myself, that I'm responsible for. It might be a rural water district that revises its long-term plan based on information my research has provided. It could be seeing something I've worked on many late nights come to fruition, like the National Climate Assessment. **Often, it's when just one person tells me sincerely that they had never cared about climate change before, or even thought it was real, but now, because of something they heard me say, they've changed their mind. That's what makes it all worthwhile, and that's what I view as success."**

For Katharine, success is also a measure of how she uses her time. "I don't plan for the future long term. I just look for what my next step is, one at a time," she says. **"To guide me, I always remind myself that the most nonrenewable resource any of us has is our time, and when we die, it won't be our academic CV that they carve on our tombstone."**

JACQUE LARRAINZAR

Mapping a city's path to racial equity

Position

Program analyst
Department of Race and Equity,
City of Oakland, California

Education

BS in music therapy and counseling
Universidad Iberoamericana,
Mexico City, Mexico

GROWING UP IN MEXICO CITY, Jacque Larrainzar wanted to be both a doctor and an artist. And she believes that her current position as program analyst for the City of Oakland Department of Race and Equity reconciles both of those early interests. **"The work I do is a form of healing," Jacque says, "and, in a way, is also an artistic endeavor that requires creativity and flexibility to see things from perspectives that are very different from my own."** Most of Jacque's work centers on critical race theory and radical inclusive engagement, while her job is to provide technical assistance to staff and departments using innovative tools and technologies, including GIS, to advance Oakland's goal of creating a community in which equity in opportunity exists for everyone.

Human rights are Jacque's passion. She knew from early childhood, she says, that she was "queer and gender nonconforming" and, as a result, faced sexism, homophobia, and repression growing up. "School was not fun for me," Jacque says. "I experienced a lot of bullying due to my appearance. I was not a good student, but I always had good grades and enjoyed learning, reading, and getting into trouble." Wanting the same freedoms and privileges that men enjoyed in Mexican culture, Jacque became involved in organizing the lesbian, gay, bisexual, transgender, and queer (LGBTQ), human rights, and women's rights movements.

Eventually, the hostile environment for LGBTQ activists in Mexico forced her into exile. She arrived in Seattle, she says, "with 20 bucks and my guitar." Reflecting on that experience, Jacque says, **"For a long time, I was not able to truly belong or be myself, but I have been lucky to find amazing people along the way who supported my dreams and helped me achieve them. If I have learned something, it is that if you keep doing what you love, you will eventually find your place."**

Changing the world

For Jacque, gender and race have always been important aspects of who she is in the world, and since she did not like the world she was experiencing, she wanted to change it. **"When I was about four, I was told I could not play with a friend that had dark skin. I could not understand why that made any difference, and the sadness of losing my friend really marked me,"** she recalls. **"As I grew up and I started to understand better the impacts of race and racism, I decided to work to create a world where my friend and I could have stayed friends for life, regardless of the color of our skin."**

Jacque is the first lesbian from Mexico to win political asylum to the United States based on sexual orientation. Before she managed to escape, she was arrested in Mexico City and tortured, an experience that left her with lingering post-traumatic stress syndrome. The most difficult moment she faced was deciding to apply for asylum, knowing that she would never be able to go back to Mexico or see her family again. But, she says, **"also knowing that thousands of people have benefited from my case and have been able to find safety, freedom, and love is more than I could have ever hoped for."**

Her experience with the immigration system led her to try to understand how institutions work, how social movements develop,

Jacque is the program analyst for the City of Oakland Department of Race and Equity.

and how change happens. She thought that studying law was the way to go but soon found that policy was a better fit. Jacque started to work on anti-racism in 2000 in the City of Seattle Office for Civil Rights, running the City Talks: Dialogues on Race, a challenging experience that allowed her to grow personally and professionally. She worked with an advisory board that included Indigenous, Black, White, Asian, LGBTQ, and straight people, as well as people with disabilities, and together they created what later became the foundation for the first racial equity initiative in the country. The program became a national best practice and has been replicated in many places.

"I am very grateful to three Black women in particular who got me started on this path," Jacque says. "Judith Vega, who introduced me to advanced race theory and the People's Institute for Survival and Beyond; Germaine Covington, the director of the Seattle Office for Civil Rights, who taught me about power dynamics in government, policy, and strategy; and Darlene Flynn, who has been an accomplice in working to transform government, first in Seattle and now in Oakland, California. Their generosity in sharing with me their lived experiences and struggles with racism allowed me to see things I was not able to understand because my experience is different from theirs."

Jacque started as a policy analyst and advanced through a series of promotions to become the policy director for the Seattle Office for Civil Rights. Part of her work was to provide support to five commissions that worked on women's issues, human rights, LGBTQ issues, immigrant and refugee affairs, and disabilities. She is proud of the work they did to protect LGBTQ people from hate crimes and to conduct the first LGBTQ census in a city. Other accomplishments included creating a domestic violence policy for the City of Seattle; decriminalizing breastfeeding and offering paid sick leave; and helping pass protections against discrimination in employment and housing for LGBTQ people, people with

Fun facts!

"I am a musician and had a rock band until I moved to Oakland."

Surprising/fun thing on her desk: "Lots of slime. I find it great to manage stress."

Tip!

"We are only bad at those things we do not try or we do not fail at enough. Approach science as a video game. It can be fun, even if you fail over and over again—eventually you will get better at it."

criminal convictions, and those receiving Section 8 housing choice vouchers.

Jacque then left the City of Seattle to work on a behavioral health survey of the needs of LGBTQ immigrants and refugees in King County, Washington, that resulted in the creation of programs to serve their needs. She also worked to create the Voter Education Fund in King County to remove barriers to voting for communities that have been historically disenfranchised. But, she says, **"When I heard that Oakland was about to start the first Department of Race and Equity in the country, I decided I wanted to be part of it, and I applied for the job to see if I could support the city in achieving its vision to embed racial equity in everything they do. After three years in the job, I can say I am very proud of what we have been able to achieve so far. Working for the people of Oakland is what I am the most proud of at the moment."**

Mapping racial inequity

Jacque's work involves supporting GIS specialists in city departments to ensure that their projects serve the goal of racial equity. GIS is essential to this undertaking, Jacque says, because geography has played a huge role in the way racial disparities affect Oaklanders, especially in zoning and planning codes but also in transportation. Her role as an analyst involves looking at what data is needed to map racial inequities and create a clearer narrative about their impact, as well as frame ways to address them. Jacque says that the work is more of a collective and team approach than an individual effort but that she provides technical assistance to keep improving the tools they've developed.

Her favorite part of the job, she says, is the people she gets to work with, as well as learning about the many different aspects of running a city. On the other hand, her least favorite part, she says,

Fun facts!

Favorite trip she has taken: "Exploring the Amazon in Ecuador."

Superpower she wishes she had: "The power to clean pollution."

What boosts her well-being: "Running."

is "dealing with the outcomes of structural racism and how we have normalized it in our everyday lives."

Along the way, many people told Jacque that working on racial equity would only bring her career to a dead end in government. But one of her supervisors, Dale Tiffany, Native American, Yakima Indian, encouraged her to become an expert on racial equity and to follow her passion for inclusion and social justice. "He gave me a card that read 'Follow Your Heart,' and I did," Jacque says.

Jacque does not give up easily; she sets goals for herself and works hard to achieve them. She also works hard to support other LGBTQ immigrants and refugees through her involvement with AsylumConnect and the LGBTQ Refugee Congress. Especially in difficult times, she says, she is motivated by hope. For Jacque, who has been shaped by her journey as an asylum seeker, success is living a life that makes the world around her "a little better, a little happier, more beautiful, and more peaceful."

Jacque's work involves supporting GIS specialists in city departments to ensure that their projects serve the goal of racial equity.

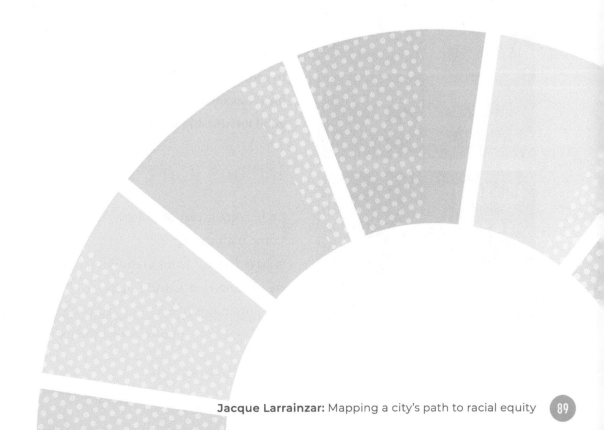

ANNITA LUCCHESI

Carving out space for Indigenous mapping

Position

Pursuing PhD
School of Geography,
Development & Environment,
University of Arizona

**Indigenous Data Sovereignty
Doctoral Scholar**
University of Arizona's Native
Nations Institute

**Board member of Gender-Based
Violence Consortium**
University of Arizona

Education

MA in American studies
Washington State University

BA in geography
University of California, Berkeley

S A FRESHMAN in college, Annita Lucchesi needed a few more units to fill out her course schedule. A friend in her dorm was taking Geography 101 and recommended the class because she liked the professor. Annita signed up for it on a whim—and fell in love. "Geography felt like the first discipline that spoke to my experiences and what I was trying to make sense of at the time," she says. "I was 17 when I started college and moved to the San Francisco Bay Area from a small rural area; I had no idea about city life, didn't feel like I belonged, and was trying to understand the economic, social, and political dynamics" in that new setting. Also, for Annita, "**Geography was the first discipline where I had professors who went out of their way to encourage me to speak up in class, or to tell me that I was smart and capable. That meant everything and speaks to how important good mentorship is for underrepresented students.**"

'Indians don't make maps'

In her senior year of college, Annita fell in love again—this time with cartography. Annita considers herself fortunate to have learned from Darin Jensen (now one of the founders of Guerrilla Cartography), who taught her the political power and freedom of expression in cartography. She was also studying postcolonial theory at the time and says that learning to draw her own maps, while immersing herself in postcolonial scholarship, was a formative experience that left her with "a deep respect for maps and the potential and intention they can have." Later, in graduate school, a professor's comment about Indigenous people not making maps inspired one of her first journal articles, titled "Indians Don't Make Maps." Annita published that article, she says, "so that the next time someone said that to me, I could cite myself proving otherwise in the literature."

During her graduate education, Annita says she has faced discrimination, a hostile environment, and at times, outright abuse. She describes graduate school as "a lonely experience," but also one that has taught her "the weaknesses of academia and the work to be done to carve out space for Indigenous intellectual brilliance (inside and outside the academy)." Despite the obstacles she's faced, she says, **"I chose to rise above those experiences because I am too stubborn to quit, because my people need me to fight for the qualifications I've earned, and because the bigger picture is all the people I can help by overcoming the challenges."**

Annita started her love affair with geography and cartography by accident and now uses it to help people, to change narratives, and to improve the future.

Mapping for liberation and justice

Though she has a long list of accomplishments, Annita says her career path has been anything but a straight line. She has worked in anti-violence advocacy, direct services, academia, and education. What all these jobs have in common is her desire to serve her community and to do work that helps drive social change. Her academic research interests include Indigenous research methodologies, Indigenous data sovereignty, Indigenous mapping practices, postcolonial geographies, critical cartography, and mapping for social change. **"My current research explores the relationships between data, mapping, (de)colonization, and violence against Indigenous women and girls,"** Annita explains. **"I'm also keenly interested in how colonial states engage in what I call data**

Annita's current research explores the relationships between data, mapping, (de)colonization, and violence against Indigenous women and girls.

terrorism, the kinds of geographies those practices produce, and what resistance to them can look like." She wants maps to be "part of liberation and justice," she says, "rather than part of maintaining colonialism."

Annita serves as founding executive director of Sovereign Bodies Institute (SBI), a nonprofit research center addressing gender and sexual violence against Indigenous peoples. SBI is home to one of the largest databases on missing and murdered Indigenous women, girls, and LGBTQ2 people throughout the Americas. In addition to the database, SBI supports community-based Indigenous researchers, is presently undertaking two international projects, and provides direct services to families of missing and murdered Indigenous people as well as to Indigenous survivors of violence. Using GIS, Annita started the database over five years ago as a grassroots community member and as a survivor of violence herself. In 2019, SBI was launched as the new home for the database and all the projects stemming from it.

Having survived traumatic gender violence, Annita wanted to put it all behind her and move on, but, she says, "I found myself drawn to work that helped me make sense of it. I wrote my MA thesis on that violence and started the database with a hand still broken from abuse, in a coffee shop at the mall because I couldn't afford Wi-Fi at home. **I had no idea that would be my life's work at the time but continued with the path put in front of me, and when I started being asked to be a voice for other survivors and stolen sisters, I felt I had to rise to the challenge to serve the people to the best of my ability.**"

Annita is also passionate about Indigenous self-determination. She has been part of mapping projects to combat extractive industries on Indigenous territories for nearly 10 years. Last year, for example, she testified against the Keystone XL pipeline in South Dakota and traveled to support the efforts of the Kanaka Maoli people to defend Mauna Kea in Hawaii.

In her academic work, Annita says, **"I have the privilege of walking the path laid out for me by other Indigenous women geographers and carving it out to be a little bit bigger for the next generation."** In her professional work, she has "the luxury of supporting those dreaming up what healing, justice, and safety looks like, and the honor of working alongside them in the trenches." In both of those roles, Annita sees herself as part of "changing who we acknowledge as knowledge producers, experts, researchers, cartographers, changemakers, and leaders." What motivates her is the opportunity to uplift others, change the narrative, and imagine new futures. "That is the work I get to do every day," she says. "It's exhilarating, inspiring, exhausting, and the most beautiful thing I know to do."

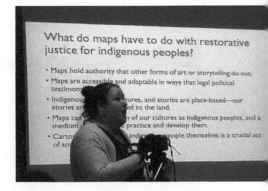

Annita serves as founding executive director of Sovereign Bodies Institute, which addresses gender and sexual violence against Indigenous peoples.

Tip!

"Don't be afraid to explore and expand your horizons. As long as you are doing your 'heartwork,' you're on the right journey."

Annita walks the path laid out for her by other Indigenous women geographers and carves it out to be bigger for the next generation.

Fighting burnout, finding strength

Along with this passionate commitment, though, comes what Annita calls "an unhealthy workaholic lifestyle" and the ever-present threat of burnout. "Academia can bleed you dry," she says. "Community organizing can do the same. I know I'm not alone in that experience, but it can sure make you feel alone. **I have had to work very hard to learn how to balance my responsibilities, obligations, and mental and physical well-being in the midst of being called to do so much. I'm still figuring it out and suspect I will be for many years.**" What has been most helpful to her in addressing these challenges, she's found, is collaborative work and the support from colleagues who create opportunities for her to take care of her mind, body, and spirit.

Annita believes that her greatest strength is the heart that she puts into her work. "Anyone can learn how to make maps, analyze data, or write a research paper," she points out. But the "radical openness, deep personal connection, compassion, and intention" in her work is what sets her apart and encourages others. And for young women considering a career in GIS, she offers these words of encouragement: **"People will tell you it's a boys' club. Don't believe them. Don't let them make you think for one second that you don't belong. There are amazing women geographers—there have been for decades, and you are part of that story."**

On a personal level, Annita feels fortunate to have deep knowledge of her Cheyenne and Italian family history. "When I feel lost or overwhelmed, I look to them and their stories and find the encouragement I need," she says. "On my father's side, I come from loggers, mill workers, and carpenters. My mother is an artist who volunteered as an advocate for abused children; my great-grandmother was a social worker. **Those backgrounds combined taught me to have deep respect for the land and environmental stewardship, to have a keen spatial awareness, to have a strong work ethic, to be community minded, and to see the power and healing potential in art. Taking that into account, it shouldn't be surprising that I ended up a cartographer and geographer, or that I wanted to use that training to serve my community.**"

Grounding herself in this history of resilience, strength, and courage, Annita reminds herself when faced with challenges that all her work is for the ancestors that came before her and for the generations yet to come. **"Everything that I do, I do for my people,"** she says. **"And by 'my people' I mean my tribe, Indigenous peoples, and survivors of violence."** For Annita, in her quest to map pathways to healing and justice, "Success is any moment where I see my people benefiting from my work, big or small." ◢

Annita works hard to balance her responsibilities, obligations, and mental and physical well-being while working on behalf of others.

SAVANNA NAGORSKI & MELISSA K. SCHUTTEN

Supporting urban development and tribal communities

Savanna

Position

Land-use planner and GIS analyst
ESM Consulting Engineers

Education

MS in geospatial technologies
University of Washington, Tacoma

BA in sustainable urban development
University of Washington, Tacoma

SAVANNA NAGORSKI and Melissa K. (Watkinson) Schutten are a unique pair—they're identical twins who both work in GIS. Savanna works as a land-use planner and GIS analyst for an engineering consulting firm, whereas Melissa is a social scientist at Washington Sea Grant (WSG), a research institute, focusing on equitable access to marine environments. The many projects Savanna is working on include developing an interactive web map for a local tribe's storm water management system and a web map study tool for local firefighter recruits. Melissa is currently piloting an urban marine program that addresses disparities in access to marine foods and resources for communities of color in Tacoma, Washington. Although they are twins, both with careers in GIS, each has charted a different path to that destination.

Citizens of the Chickasaw Nation, Savanna and Melissa grew up with family enrolled with the Upper Skagit Indian Tribe. Melissa says it's no surprise that she and her sister have both worked on environmental issues since their favorite memories are of being outdoors—camping, fishing, and crabbing. For most of their childhood, their dad worked for tribal housing services with different tribes in Washington. "I believe that our own Native heritage, and growing

Savanna, *left*, and Melissa with Savanna's twin daughters, Dahlia, *left*, and Coral.

up around Native communities, instilled in us the value of cultivating strong relationships and working with tribes," Melissa explains. "We also grew up in a household that always had someone outside our immediate family living with us, or we were often doing activities that supported the community in one form or another. This nurtured altruism that is centered around collective strength and giving all that we are able to support others."

Going their own way

At 18, the sisters had their first real experience of being apart when Melissa moved to Seattle to attend Seattle Central College, where she completed her associate of arts degree. "We both really desired some independence, and I'd say we both took that to heart in our own ways," she says. "Those first few years out of the house and on my own were pretty transformational in learning and participating in urban communities."

"I would say that was when we really started to appreciate and miss each other," adds Savanna, who attended university for a semester before returning home for community college. The two were in and out of college for a few years while they figured out what they wanted to do, before landing at their respective University of Washington schools and programs. By chance, they happened to pick up GIS in their different fields of study.

Melissa

Position

Equity, access, and community engagement lead
Washington Sea Grant

Education

MA in policy studies
University of Washington, Bothell

BA in global studies and society, ethics, and human behavior
University of Washington, Bothell

Melissa clamming for butter clams near Port Gamble, Washington.

Discovering a calling

Savanna discovered her passion for urban planning and community development when she was 20, living in San Francisco and volunteering with Youth With A Mission (YWAM). "The diversity of residents, rich and poor, all need a seat at the table. During my time in San Francisco, I engaged with the homeless population and learned about how they managed their own lives in hardship but also made the most of opportunities in the community, such as fishing on the local piers, participating in church activities,

Fun fact!

Favorite trip of Savanna's: "We went on a road trip together in 2015 from Oklahoma to Arizona. Our main priority was to visit our tribe, the Chickasaw Nation, to show gratitude to the educational department that afforded us our higher education degrees and to visit our tribe's cultural center. Chickasaw also has a state-of-the-art geospatial technologies department. We visited their offices to meet the team and were gifted beautiful maps the tribe produced. We've got a few of them framed and hanging in our homes."

Melissa, *left*, and Savanna on their Chickasaw road trip in 2015.

and assisting vendors at the local farmers' markets," she says. Later, when Savanna started researching college paths, she came across the Sustainable Urban Development degree in the urban studies program at the University of Washington, Tacoma (UWT).

In 2009, Melissa transferred to the University of Washington, Bothell (UWB), where she would complete both her undergraduate and graduate degrees. But, she says, **"If there is one thing I've learned that is certain, it's that there is no one direct or single path for someone to get where they are today or where they want to go in the future."** When she first considered going back to college after completing her associate's degree, she looked for programs focused on communications. She and Savanna had an interest in becoming journalists or photographers, something they attribute to their mother, who worked at a local newspaper when they were young and enjoyed photography. When the communications program didn't turn out to be what Melissa was looking for, she enrolled in a double major: society, ethics, and human behavior, plus global studies.

But in the two years between graduating with her bachelor's and beginning her master's, a significant event caused Melissa to change direction: "My grandmother passed away, and while my memories of her are few, she did make an impact on me when reminding me that 'our own people need help, too.' I pivoted from pursuing an international career to one that is hyper local—working with Native American tribal communities." Through a public lecture, Melissa also learned about the critical and urgent effect that climate change was having on many coastal communities of color, including Indigenous communities. Ultimately, she says, she applied these lessons by developing qualitative, quantitative, and spatial analysis skills in her graduate program through "partnering with a coastal tribe to identify the impact of historical land policies on the tribe's capacity to adapt to climate impacts."

Making an impact with GIS

The studies of one of Melissa's advisers at UWB, Dr. Santiago Lopez, aligned with her interests in working with Indigenous communities on climate change. From an Indigenous community in Ecuador where he conducts research on climate change, Dr. Lopez uses GIS in much of his work and was the primary GIS instructor on campus at the time. For Melissa, taking his GIS courses was a no-brainer: she believed that GIS would be a skill that could help advance her career and that she could use when collaborating on future research projects. Also, she explains, **"My attention and understanding of the world works in a very spatial way."**

Though she struggled with math and science in high school because of a lack of support from teachers, she found a love—and

Melissa has worked on several projects in partnership with local Native American tribes to identify the socioeconomic and community impacts and resilience factors of ocean changes.

Credit: © Sharon H Chang

talent—for numbers, statistics, and analysis during graduate school. But beyond the university being a place to learn and strengthen her skills, the most important part of her schooling, according to Melissa, was meeting the people closest to her, including her wife, Cyra.

For her capstone project at UWB, she partnered with a Washington coastal tribe to conduct a study of the impacts of Indian allotment and assimilation policies—which have led to fractionated and checkerboarded lands on many tribal reservations—on a tribal community's ability to adapt to climate change. The tribe that she worked with had begun planning to relocate much of its community infrastructure because of impacts to its village such as land erosion, increased flooding, sea level rise, and risk of tsunamis. "Because these cumulative effects were forcing the tribe to consider relocating to higher elevations, the tribe would have to buy back parcels of land that had remained outside the tribe's trust as a result of these allotment and assimilation policies," Melissa says.

Using GIS, Melissa helped identify which parcels of land that remained out of trust the tribe should consider buying back. "Throughout this research project, I applied a community-based participatory research model, which I had learned about during my first graduate research position at the Indigenous Wellness Research Institute," she says. "This model encourages the recognition of autonomy and organizes the researcher to work in partnership and as a co-collaborator with communities. I continue to use this model in most of the research projects I participate in, as it respects and acknowledges the sovereignty and self-determination of communities, and particularly tribal nations." In her last year of graduate school, Melissa held another graduate research position at WSG, where she used different methods, including GIS, to help develop the Socioeconomic Assessment for the Washington State Marine Spatial Plan.

Melissa, *left*, with her wife, Cyra, at a UW game.

Tip!

Melissa: "The world will be a better place when your knowledge, experiences, and perspectives are reflected in STEM and GIS."

Dahlia and Coral as babies wearing UW T-shirts.

Succeeding despite it all

When Savanna started her undergraduate degree at UWT, she was 23 and a newlywed. Then, during her second quarter, she found out she was pregnant—with fraternal twins. "As a military wife battling a tough pregnancy, keeping up with my schoolwork was a challenge," she says. "My professors never gave me a free pass, but they were very accommodating and flexible with deadlines, which helped me continue through my studies. I took a single quarter off school after having our babies and went back full time. With the help of family and friends to help care for our babies and even clean my house, we made it through." She even found time to participate in extracurricular projects, such as contributing to the publication of a paper on GIS and urban development practices that earned her a student researcher award.

Savanna credits several women with giving her the motivation to keep going. When she was struggling with managing schoolwork, being a military wife, and being a new mom to twins, one of her professors, Dr. Anne Taufen, reminded her that "80 percent of success is just showing up." "This gave me the freedom to give it my all 20 percent of the time," Savanna says, "and helped me to successfully move forward in my educational pathway." Another professor from her MS program, Dr. Britta Ricker, opened up about her own struggles throughout her career, encouraging Savanna not to give up. Her local chapter of Women in GIS and Technology, led by Tonya Kauhi, has also been supportive, helping Savanna learn more GIS skills, establish her career, and connect with other women. **"I won't pretend that it's always easy to keep going, but I understand the power in continuing to move forward even with adversity,"** Savanna says.

Savanna in front of her GIS certificate capstone project in 2014.

Savanna with Dahlia, *left*, and Coral at Savanna's MS graduation in 2016.

Improving the community with GIS

Savanna was introduced to GIS during her urban studies degree. Her first class, Beginning GIS, was challenging and she wasn't particularly interested in it at the time, but she knew that having a GIS certificate would be beneficial to a career in urban planning. After she graduated, she worked in a local GIS and planning internship position, where she explored applying GIS to real-world problems. **"This position was eye-opening in that I realized I thoroughly enjoy the GIS field, while also stirring a passion in me to work within the community in urban development."**

During her MS program, Savanna worked as the GIS coordinator at UWT, which provided several unique opportunities to expand her skills. "One example is a student housing analysis that helped to establish evidence that additional student housing was needed on campus, which then led to the school purchasing apartment units for the current enrollment and future growth," she explains.

Melissa, *right*, with her wife, Cyra, and their two dogs.

Her GIS certificate capstone project studied locating potential neighborhood gardens around eateries in the business district of Tacoma that had a likelihood of creating food-sharing programs. "The following year, my family converted our front yard to a neighborhood garden. We were able to build community within our neighborhood, participate in the city's community garden program, and share the harvest with our neighbors," she explains. Then, for her MS project, she partnered with Harvest Pierce County to develop a beta Harvest Share app, in which ArcGIS Online apps were integrated into a mobile-rendering website for community gardeners to share excess food with their neighbors.

Working toward a better future

While Savanna conquered her schooling, Melissa went from graduate research assistant at WSG to marine policy fellow at The Nature Conservancy (TNC). TNC already had two in-house GIS staff, and Melissa partnered with them on several projects. For example, she developed an Introduction to Tribal Engagement training after recognizing that there was limited understanding of the intersection between tribal lands and TNC's conservation lands. "Tribes and Indigenous peoples have been stewards of these lands and waters since time immemorial," Melissa says. "I believed it was important that, as an organization conserving lands and waters that had been stewarded by Indigenous peoples, there was a deeper understanding of the Indigenous people's history and relationship with the land, particularly as environmental organizations aim to build stronger relationships with tribes." Melissa partnered with TNC's GIS staff to create a map that reflected TNC's conservation lands and the tribal reservations of Washington State. "It was eye-opening for many of the staff, who hadn't considered how their work directly relates to the histories and lives of Indigenous peoples," she says.

Melissa landed her first full-time, permanent position as a social scientist at WSG. Her favorite part of her job, she says, is "nerding out" over data that helps build equity within communities. The harder part is that **"sometimes it's really difficult to access spatial data, and sometimes it takes a while to see the direct benefits of your work, so it's important to find ways to acknowledge the short-term accomplishments,"** Melissa says.

She's most proud, though, of building the conversation and influencing policies around Indigenous stewardship and the impacts of ocean changes on communities, including the factors that contribute to resilience. Her first big assignment at WSG was to support the social science aspects of a study on impacts of ocean acidification on Washington's outer coast. This project was the first transdisciplinary and multipartnered project she participated in, with principal investigators that included oceanographers, social scientists, marine biologists, and tribal nations, among others. She was again encouraged to practice community-based participatory methods, as well as to learn and apply Indigenous science.

Savanna and Melissa's preschool graduation.

Starting strong

Originally hired at ESM Consulting Engineers as an assistant planner and GIS technician, Savanna immediately proved her GIS skills and her title was changed to GIS analyst. **"My GIS skills are primarily used during the feasibility stages of a project, where the planning aspect of my work is significantly enhanced by having access to data that several other companies may overlook or do not consider.** I've also had the opportunity to produce web maps for developers and agencies that otherwise would only be able to view their data in as-built plans," she says. Her favorite part of her job is the diversity in work opportunities, especially being able to help build the foundation for land development projects within her region.

Savanna and Melissa at one year old.

Although Savanna quickly proved her abilities at her new job, she struggled to get that opportunity in the first place. She had a hard time finding a job, interviewing for dozens of positions while her peers had no trouble getting hired. "I genuinely felt that because I am a woman with children, particularly when my husband was in the military, it was far more difficult to be seen as a great candidate," she says.

Striving for an equitable world

This perceived bias against being a mother, in addition to the challenges she overcame during schooling, has made Savanna passionate about encouraging young people, especially young women, to explore and be proactive with their education, career, and life opportunities. Through IGNITE Worldwide, she's participated in panels to share her experiences with middle school and high school girls in her region. She also enjoys being a panelist or participating in social engagements with other UWT urban studies students.

Savanna with her daughters at the Esri User Conference in 2019. She displays the maps that they made in her office.

For Melissa, creating a world where future generations will thrive is what motivates her. **"I wish to be a good ancestor. For me, believing that I can create a better future for my nieces is my greatest motivator,"** she says. "What makes me happy is to have a healthy and secure family, to belong to a community, and to know that my work has a positive impact on others." She credits Vi Hilbert (Lushootseed name taqʷšəblu), her dad's aunt, a respected elder of the Upper Skagit tribe, with leaving a legacy that inspires her and her work every day. "The imprint that she left across the region provides regular and invaluable reminders of her leadership and commitment to ensure that Native peoples and culture were sustained and honored," Melissa says.

Although she has strived to create a sustainable world, she has also struggled with her place in it. "In most stages of my

Melissa with her sign at a women's march.

educational and professional career, I had to overcome what is now more commonly known as 'impostor syndrome,'" she says. **"As a gay, Native woman, I did not see others that looked like or reflected me in most stages of my path. I had to work to acknowledge that I am in the right place, I am more than capable of having success here, and by my being here, I can add to the representation."** Savanna, too, has faced impostor syndrome in her job, but says, "I have over-come my own weaknesses by having candid conversations with other women in the industry and with my male bosses." Yet recognizing these challenges, and confronting them, has strengthened Melissa's sense of her purpose, which she says is "to work toward diversity, equity, and inclusion. **We all gain in a world that is more diverse, where everyone has equitable access to resources and decision-making, and where all people feel that they are included in the places they wish to belong."**

Tip!

Melissa: "To me, when you can find joy and happiness, you have found success. Success is not a destination, it is a journey. In recognizing this, each person can experience success, and it will look entirely different from one person to the next."

TRISALYN NELSON

Turning geography into practical solutions

 HEN DR. TRISALYN NELSON dropped out of all her classes in the first term of her second year at university, she could never have imagined that today she would be a professor—the Jack and Laura Dangermond Chair of Geography at the University of California (UC), Santa Barbara. "I am a first-generation student," she says. "When I first started my undergraduate, I literally couldn't find the classrooms. There were many things I didn't understand about how to navigate university, and in retrospect, I wasn't academically prepared. I had a lot of great support from friends, family, and professors that got me through. By the third year of my undergraduate degree, I had figured out how to be a student and had more fun and more success."

Solving problems, keeping cyclists safe

Today, Trisalyn and her team at UC Santa Barbara undertake collaborative projects that use spatial methods to solve new problems: "I have studied mountain pine beetle infestations, grizzly bears, and environmental change, but currently my work focuses on active transportation and the use of big data and analytics to better plan cities. I am really proud to be the founder of BikeMaps.org, a web map and app to gather volunteered geographic information on

Trisalyn bicycles for a healthier lifestyle.

cycling collisions and near misses." Cities have invested in safer infrastructure because of the data the group has collected, she says. "It is hard to quantify the impact, but very likely someone avoided injury because of the BikeMaps.org project, and that is amazing," she adds.

Trisalyn emphasizes that she is "passionate about the link between happiness and physical movement. The less time I spend in my car, the happier and healthier I am. Many of the projects I work on have the goal of making active transportation safer and therefore more accessible to a wide range of people."

Learning by doing

Growing up in Victoria, Canada, Trisalyn was the second oldest of five children. For work, her mom, who was single, took in six mentally handicapped adults that she cared for in their home.

"Basically, it was my mom and 11 dependents. As one of the older children, I had a lot of responsibility and often took on a leadership role in the house," she says. "These skills translated into good life skills: work ethic, time management, and finding creative solutions. Though in many ways I was not academically prepared when I arrived at university, the skills I learned in my family helped me succeed."

Trisalyn says that she learns by doing and that the hands-on work of internships was key to her academic success. She was also happy to find a mentor who taught her that "forest conservation, my passion as an undergraduate, was something I could positively influence by applying my GIScience skills to support better decision-making," she says.

Although, among other prestigious positions, she was the first woman to be a full professor in the Department of Geography at the University of Victoria, Trisalyn admits to a recurring nightmare that she hasn't finished her PhD. "There is usually a course that I forgot to take, and my degree isn't valid," she says. "If that is not a sign of impostor syndrome, I don't know what is. **Academia is a tough place to believe you belong. While I love it and am often confident, I can still battle the feeling that I am not a 'real' academic." But, she jokes, "then I just put on a jacket with elbow patches for a day or so."**

Trisalyn loves bicycling with her family—husband Ian Walker, also a geography professor, and their children, Beatrice and Finn.

Fun fact!

"I have a line of GIS baking. I love to nerd out and bake sugar cookies that look like Voronoi polygons. Chocolate chips in cookies are the results of a random spatial process. Yummy point patterns."

Trisalyn uses GIS in her bakery designs.

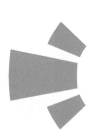

Opening doors

As a professor, Trisalyn says she is passionate about diversifying GIScience, working with colleagues to develop new pathways into the field for underrepresented groups. **"A lot of young women scholars I have met (including myself) talk about wanting to do work that has a positive impact on people or the planet,"** she says. "At times, academia can seem very theoretical, and I think that can be a deterrent to some young women. I think you can do so much great and applied work in academia with GIS. **The key is to find good partners and learn to communicate with people outside of GIS."**

Trisalyn has two children, ages 9 and 12, and her husband is also a geography professor. Balancing family and work is always on her mind. "One of the ways I balance work and family is to engage my family in my work," she says. "I have had my children on my podcast and get them to help in outreach for BikeMaps.org. Geography is something most kids love because it is about people and place." Trisalyn has written articles on balancing academia with parenting and uses all kinds of tricks, from cutting corners on household chores such as folding laundry to optimizing how you use small amounts of time efficiently.

In her four years as director of the School of Geographical Sciences and Urban Planning at Arizona State University, she led the school in hiring 13 faculty, doubling research revenue, generating $2 million in new revenue, increasing diversity of majors (more than a threefold increase in minority students), and growing undergraduate majors by 25 percent. Trisalyn values partnerships, particularly connections between industry and academics, that enable innovating methods and approaches for solving critical issues. She has worked with more than a dozen partners to apply data science methods to business and management solutions.

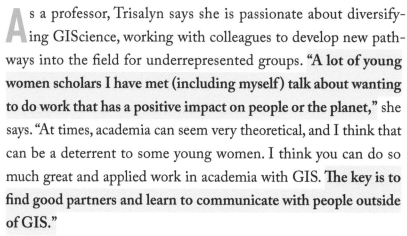

Fun facts!

What she wanted to be when she grew up: "A band teacher. My dad owned a music shop, Nelson's Music. But it turns out I have no musical talent."

Superpower she wishes she had: "I wish I could teleport. Teleportation would remove all the geographic constraints in my life. No more distance between me and the people I love or between me and adventure. Plus, it would be superefficient. And I could think about spatial autocorrelation in new ways."

"Something I love about GIS and STEM is that you have real skills. Having skills is very empowering. Someone brings me data, and I have the skills to turn it into new knowledge. That is a cool feeling. Like Willow in *Buffy the Vampire Slayer* – the early seasons."

One of her proudest accomplishments, she says, is that she has helped students grow and succeed in their studies. **"It is awesome to see a student gain skills and confidence, and then head out into the world to generate GIS solutions,"** she says. With a firm belief in the forceful solutions that geographers can provide the world, Trisalyn is excited about educating the next group of GIS professionals and scholars. "I really love working with a team of students and collaborators that bring unique skills together to solve problems," she says. **"The problem might be where to put bike infrastructure or what roads to close to reduce grizzly bear mortality, but when you bring together a team that has diverse skills, it is possible to find real solutions."**

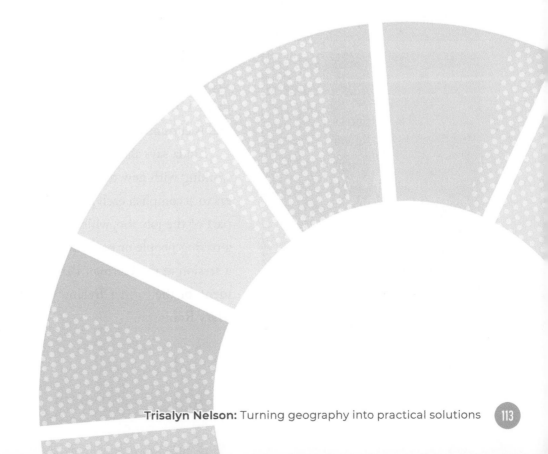

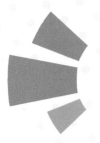

LINDA OCHWADA

Leading the way on geospatial AI and innovation in Africa

LINDA OCHWADA'S PERSONAL motto is "with problems, create alternative solutions." It has served her well as the managing director of AfroAI and CEO of FIKO Tech. AfroAI consults in machine learning and artificial intelligence on geodata, with a focus on the African market and European companies that work with Africa. FIKO Tech is a proximity alert and security system that combines Global Navigation Satellite System (GNSS) and cellular technology to increase security. In cities in developing countries, especially those with poor power and electricity infrastructure, the process of getting through a secure gate (such as at home or in an office) can be time-consuming and even dangerous. FIKO Tech solves this problem by transmitting a person's impending arrival to the security guard or family members to alert them. Linda's also interested in data-driven climate research and is currently applying for PhD positions in that area.

Linda says she enjoys every part of the work she does—working with new technology and the knowledge she gathers to accomplish each project or task. "I like the networking part of the job, too, which gives me the chance of meeting amazing people or participating in great events like having a session at the German Parliament, meeting the Esri president, or the former Ireland president, and my role model, Mary Robinson."

Linda presents a project at Geek Girl Carrots' "Hack Like a Girl" hackathon in 2018.

In addition to technology and innovation, Linda is passionate about feminism (women's rights, especially in technology) and about combating injustice (for women and the LGBTQ community). "Women having their own source of income drove me into starting Kenderaalala Women Group (KAWG). This is a community-based organization with an aim of increasing the income and livelihood of women through environmentally friendly agriculture activities in the Samia District in Kenya," she says.

"Most technological solutions are from a man's point of view, how men perceive a problem," Linda says. "For example, when I presented FIKO technology to a group I am in, one question that came across was why couldn't I use a siren at the gate. Eliminating the danger possibility on alerting thieves to one's arrival, I considered the noise pollution that hooting and a siren cause. As a mother, my son being woken up by my neighbor coming home with his car in the late hours is a problem for me."

Striving for a better future

From a small village in Kenya, Linda says, "having a backpack or shoes made you feel out of place. Everyone was from a humble background." But driven with ambition from a young age, Linda was determined to achieve big things. "I am really target driven. I was always giving myself a target and would hit the target, and then make a new target," she explains. In Kenya, the school determines a child's next level of studies, so rather than get her grade certificate from her small village school, she moved to an urban area to stay with her aunt to attend a better school. "With the same ambition, I would move to new targets, one by one," she says.

As a child, Linda says she was the student to look up to, mentoring others in math and science. She originally intended to become an aeronautical engineer but ended up focusing on geography at university. During her undergraduate studies, Linda says she tried to not be called the *choppy sheng* (Kenyan slang for "bookworm") and decided to relax and try some extracurricular activities. One such extracurricular activity was modeling, which sparked an interest in fashion for Linda. **"I started modeling and fashion to prove a point in Kenya, that you can be bright in school and also do fashion—that there was no negative correlation between fashion and brains,"** she explains. She's even launching a fashion house in 2021 called Eburi, which is Afro-Euro fusion and aimed at highlighting the beauty of those with Black backgrounds, Linda says.

Coming out on top

For her graduate degree, Linda faced the hardest decision she's had to make. She moved thousands of miles away from her family in Kenya to Berlin, Germany. "I started my master's in geodesy and geoinformation science when my son was only five months old. In a new country, with a kid, by then staying alone, transitioning from a social science to an informatic course and with no programming skills—at that time, when I couldn't keep up with all

Linda hiking with her son.

Fun fact!

"I like fashion. In fact, I have done some beauty pageants, runways (including Berlin Alternative fashion week), and most of all, I love sewing and designing clothes— I have done some sewing this year."

Linda modeling at a fashion show.

Linda says the best advice she was given when starting out was "live today and make sure you never regret yesterday."

the things around me, I was called a 'weak link.' That really brought me down, but not to an extent of discouraging or giving up my passion," Linda explains. In fact, Linda took on an add-on master's program—Climate KIC—after her first year. Knowledge and Innovation Community (KIC) works to accelerate the transition to a zero-carbon economy.

"With no knowledge of German as a language, I had to come back home at 9:30 p.m. and prepare dinner for my son before doing homework. The most challenging part was my transition in courses. BA geography is more of a social science course while MSc geodesy and geoinformation science is more statistics, geometry, and programming, most of which I had a high school knowledge of or no knowledge at all. I had to learn four programming languages with applications in engineering, statistics, computer vision, and so on. At times, I would be forced to carry my son to a programming class," Linda says. Despite the hardships, Linda graduated successfully.

"I am proud of every step I made that led to my achieving my goals or attaining my target, including the hurdles I moved past," she says.

Fun fact!

Favorite trip: "A road trip from Busia to Kitale to Lodwar to Loitokitok to Taita Taveta Town to Voi to Mombasa to Nairobi. This was one of those unplanned trips that I had no hotel plans but rather jumped on a bus to a destination I wanted."

Seeing data as the future

During her undergraduate degree, she developed a passion for GIS and remote sensing that led to her starting her geospatial career working at the United Nations Food and Agriculture Organization's Somalia Water and Land Information Management Project (FAO SWALIM) as an intern. From there, she worked for Save the Elephants in Nairobi, where she monitored elephants' migratory routes and their interaction with humans by mapping and studying the elephants' routes using a GPS collar on the elephants. Her work with FAO SWALIM helped her see GIS and remote sensing as more than just a theory course. Being able to work with real data and see the impact of her work made her want to help the community—especially small-scale farmers like her mother. **"GIS is the technology of the future," she says. "We are having a globalized data economy, and to be exact, a spatial data economy."**

After graduating with her master's in Germany, Linda worked as a geodata scientist before launching AfroAI. She developed the company after noticing a gap in consultancy when she attended

Linda on a farm in Kenya.

events, especially for companies that work with African data. "Understanding the data is the first step of a useful insight to data," she explains. **"My goal is to be a major player in the Africa-European trade and corporations, especially in tech."**

Fun facts!

Fun thing on her desk: "I have figures of Fire Fist Ace and Straw Hat Luffy [manga characters] and also a painting of Fire Fist Ace that I got as a present from my sister-in-law."

On her bucket list: "Explore the Trans-Siberian Railway, and then fly from Beijing to Japan to the anime capital, Akihabara."

ZARITH PINEDA

Generating empathy through equitable design

HEN ZARITH PINEDA was seven years old, her family emigrated to the United States from Colombia, displaced by a violent territorial conflict in their homeland. Her family's decision to uproot themselves and begin life again in the US not only shaped Zarith's childhood but inspired the direction of her future career. The decision to leave one's country, she says, "is never made lightly. It comes with the unique pain of losing a physical place and irreplaceable attachment to a sense of *home*." It's a trauma that affects millions of people around the world, forcibly displaced by armed conflict, climate change, or gentrification. And in some ways, Zarith says, her work to advance spatial justice is directly related to "a desire to prevent this sense of spatial loss in other families."

Zarith is the founder and director of Territorial Empathy, a nonprofit design collective that studies urban equity issues affecting marginalized communities—women, children, and people of color. The goal of the organization, she says, is to "investigate the architecture of oppression in urban environments to propose empathy-based design recommendations that support inclusive communities." For Zarith, this means working with community members, "folks that don't usually see themselves reflected or engaged in this type of work," and taking the time to understand their perspectives. **"Whether it is through storytelling, maps, or spaces, I think we can all use our tools of practice to connect rather than divide."**

<table>
<tr><td>

Position

Founder and director
Territorial Empathy

</td></tr>
<tr><td>

Education

MS in architecture and urban design
Columbia University, New York

M.Arch in architecture
Tulane University, New Orleans, Louisiana

BA in French
Tulane University

</td></tr>
</table>

Currently, with her team, Zarith is working on Segregation Is Killing Us, an ongoing project that looks at New York City's racist planning history to explain the disparate impact of COVID-19 on communities of color. The research is presented as an interactive map telling a story that visualizes the correlations between such factors as redlining, air quality, health-care access, race, income, and COVID-19 clusters.

Architecture reimagined

Growing up, Zarith imagined that she might become a doctor, but art and architecture were an intimate part of her life from an early age. Her mother, Antinea, started architecture school when Zarith entered prekindergarten, and some of Zarith's earliest memories are of playing with her mother's building models, going to studio with her, and being surrounded by and encouraged to use her art supplies.

Zarith says that her mother is the woman she admires most, both personally and professionally. After immigrating to the US, Antinea had to restart her career from the bottom up, which meant working many jobs as a single parent and, at the same time, going back to school to validate her degree. She later started her own successful design-build firm in the Greater Boston area. "She is a respected authority in what was largely an impenetrable boys' club," Zarith says, with pride, "and has always encouraged me to persevere, no matter the obstacles."

When Zarith went to college at Tulane University in New Orleans, she majored in both French and architecture. At the time, in the aftermath of Hurricane Katrina, the architecture school was focused on the reconstruction of New Orleans, but Zarith's own focus shifted radically in 2013, when she joined a research trip to the Balkans. The purpose of the trip was to explore the impacts of *urbicide*—genocide against the built environment. Zarith says, "I'd never heard this term before, and learning about it moved me

Zarith founded the nonprofit Territorial Empathy.

Zarith, *right*, with her twin sister, Zahira, and her mother, Antinea.

deeply." As she studied further, she was introduced to the notion of cities as "systems that could be manipulated to perniciously control, violate, and subjugate people." For Zarith, "Learning that this is actually the norm and not the exception changed the trajectory of my career, which at the time had been limited to the architecture of a building and not the architecture of systems in urban environments."

'Engineered Paradises'

Inspired to explore the ideas of urbicide and spatial justice for her thesis at Tulane, Zarith came up with a topic that was "a bit of a departure from the status quo, to say the least." The thesis, "Engineered Paradises," imagines spaces of reconciliation between Palestinians and Israelis on the West Bank.

"While most professors criticized my approach, I was fortunate to have Graham Owen as my thesis adviser," she says. "To this day, he is one of the most influential educators I've encountered. He taught me that **great design has the power to suspend reality and therefore inspire people to imagine the unimaginable.** His faith and support gave me the space to explore and produce some of the work I'm proudest of." "Engineered Paradises" would go on to be deeply influential in the development of Zarith's career, receive national recognition, and continue to be published, exhibited, and referenced for years to come. "This experience taught me that **a little encouragement goes a long way,"** Zarith says, **"and it's a practice I try to implement with my team, students, and in my personal relationships."**

Another professional milestone was one of the most impactful projects of Zarith's career, the D15 Diversity Plan—the first community-led public school integration plan in New York City. School District 15 in Brooklyn is one of the most racially and socioeconomically diverse regions in the city but, at the time, had some of the most segregated schools. Working as an urban and architectural

Fun facts!

"Racism, misogyny, homophobia, xenophobia … all begin when we conveniently choose to forget another's humanity. This is why I'm passionate about exploring creative ways of generating empathy through design. Empathy makes dehumanization really hard."

Best advice she was given when starting out: "We live in an abundant world where there's enough room for everyone to coexist and thrive."

Field research in the Middle East for "Engineered Paradises."

designer for a local firm, Zarith led community engagement efforts for the project, particularly in the Latinx community, and through GIS analysis, depicted those patterns of segregation to build community support for the plan.

Empathy-based design

Although the project had a successful outcome, it was an eye-opening experience in other ways, as Zarith witnessed up close the limits of inclusivity in planning initiatives. This experience, she says, spurred her to found Territorial Empathy as a nonprofit organization not beholden to special interests but based instead on "truly meaningful and healing community engagement." Since the nonprofit's founding in late 2018, Zarith says she is proud of the work her teams have accomplished and is equally proud of the

Zarith presents "Engineered Paradises" at the American Institute of Architects conference in Washington, DC.

Zarith practices youth engagement in Poughkeepsie, New York.

"truly intersectional" nature of these teams: "I feel fortunate to work with incredibly talented and thoughtful people who demonstrate every day that working relationships can be supportive and pleasurable," she says.

At the same time, fund-raising is still one of her biggest challenges. There are many barriers to entering the circles that fund organizations such as Territorial Empathy, and the world of philanthropy, Zarith says, still suffers from systematic racism and misogyny. Yet she is hopeful that the recent reckoning with race relations in America will help improve the fund-raising prospects for her work.

In the face of all these challenges, Zarith believes that her greatest strength is faith in herself. It's a hard practice, she says, but one that has allowed her to succeed. "I can't tell you the number of times I've been counted out or underestimated. I'm a petite woman of color working at the intersection of tech, urbanism, and architecture. In the United States, POC [people of color] women are largely underrepresented in these fields," she says. **"The design of our buildings and cities has excluded a large part of the population**

"Girls, the fate of the world depends on you. It depends on your brilliance, magic, strength. This is not an exaggeration. We need you in any room where decisions are made, because if you're not there, chances are the decisions being made don't have your best interests at heart. I want you to operate at a minimum with the sense of entitlement that our society reserves for white men. This means you hold your head up, and you do whatever it is you want to do. If that's pursuing STEM, great! If this pandemic has taught us anything, it's that facts and science matter. We need you on the inside of these institutions to help the public restore the faith in their integrity."

and has been manipulated to benefit the privileged. So, when you begin inhabiting these spaces and you look around the table and no one really looks like you, it's hard to feel confident. Impostor syndrome is real. But there is no choice but to believe in yourself, your experiences, and your abilities if you want the people you represent to succeed."

As she continues to strive for her vision of spatial justice, Zarith is motivated by the young people she encounters and mentors through Territorial Empathy. "I'm moved by their resilience, adaptability, and hope for a better future," she says. "I feel accountable to them. I want to do work that supports them living in a more just world. Work that empowers them to live happy, healthy lives."

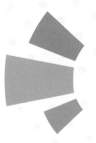

MAYA QUIÑONES

Bringing forestry data to life in the Caribbean

Position

Program manager, US Caribbean region (Puerto Rico and US Virgin Islands)
US Department of Agriculture (USDA) Forest Service, International Institute of Tropical Forestry, San Juan, Puerto Rico

Education

MSc in remote sensing, image processing, and applications
University of Dundee, Scotland

BA in geography
University of Puerto Rico, Rio Piedras

MAYA QUIÑONES IS THE FIRST to admit that she spent many hours of her childhood playing video games. Growing up in San Juan, Puerto Rico, she was raised by her mother, Sylvia Zavala Trias, a forward-thinking public school librarian, who made sacrifices so that the family could have a computer and a game console in the house.

"We didn't have a lot of money," Maya says, but "technology and books were prioritized in our home over other commodities." Maya loved playing with the Nintendo, the Apple II, and, later, the Macintosh Plus. And, she believes, far from being a distraction, her video game habit may actually have paved the way for a career in GIS: "Playing long hours of 8-bit RPGs (role-playing games) and puzzle games made it familiar to view the world in two dimensions and work with maps. So when I started learning GIS and cartography, it just made so much sense."

Today, Maya is based at the International Institute of Tropical Forestry in San Juan, Puerto Rico, a USDA Forest Service research station, where she manages two USDA Forest Service programs for the US Caribbean region: the Urban and Community Forestry Program and the Forest Stewardship Program. She is currently working on collaborative projects with the US Forest Service Geospatial Technology and Applications Center and the University of Wisconsin, Madison, to create the forestry data needed for improved urban forestry planning and management—data that includes lidar-derived canopy cover data, wildland-urban interface data, and refined island-scale urban areas for the Caribbean islands.

Before becoming a program manager, Maya worked for 15 years as a cartographic technician. She used to make maps daily, and she still misses the creative process of making a map with "the right colors, the right font and scale and labels and other elements that communicate its intended message effectively." That process, she says, is "mesmerizing" and so much fun. But, as a cartographer, she saw the need for geospatial data in the Caribbean islands.

"Our island territories have been historically left out of national datasets," she says. "This has been changing in recent years, in part because of devastating hurricane events that have kept us in the news and in people's minds and, in large part, because people are raising their voices for more inclusion and equity."

Thinking spatially

A s a child, Maya had no idea what she wanted to be when she grew up; she lived very much in the moment and in the imaginary world of video games. At school, geography lessons were uninspiring because they mainly involved memorizing capitals and rivers. And, she admits, she wasn't really interested in studying, so her high school GPA was unimpressive. But she did study hard for the entrance exam to the University of Puerto Rico, Rio Piedras, got a high score, and was admitted to the Social Sciences Department. Then, in her second year at the university, everything changed: in a class called Elements of Geography, taught by Dr. Nancy Villanueva-Colón, she was introduced to geospatial thinking. "The concept of scientifically looking at the world as it relates to the space it inhabits was new to me," she says. "But once I did, I fell in love with it. Everything has a spatial component. It's like everything is part of this big puzzle, and you can see how things

Maya loves making maps, choosing the right fonts and colors, and working with geospatial data.

Maya working on her research map, *Spatial Analysis of Puerto Rico's Terrestrial Protected Areas*, 2012.

Credit: Gary Potts.

fit together by looking at their spatial relationship. So I quickly switched majors to geography."

As a geography major, researching coastal mangrove systems for a marine geography class, Maya managed to get an interview with Dr. Ariel Lugo, director of the International Institute of Tropical Forestry. After the interview, he invited her to volunteer at the institute—which she did, and where she still works today. Back then, as a student volunteer, she was assigned to assist Dr. Eileen Helmer, a research ecologist specializing in remote sensing, and, Maya says, **"Everything I had learned in class fell short. My GIS, cartography, and remote sensing skills were born with this experience."**

Finding mentors, making maps

The two GIS specialists in the lab at the time, Olga Ramos and Dr. Tania del Mar Lopez, taught her all the basics, while Dr. Helmer taught her remote sensing, how to work with satellite images, edit raster datasets, mask clouds, classify vegetation, and so much more. "It was an amazing experience," Maya says. "It was their mentorship that allowed me to have a career in geospatial technology. **I often tell our interns that, while what you learn at university is an important foundation, you truly learn your craft on the job.** So, don't worry if you don't feel like an expert when you first apply for a job."

Today, Maya expresses deep gratitude to her mentors at the institute. "I admire all three of them," she says, "experts in their fields paving the way for the rest of us and always willing to help others. In my experience, having mentors at work that took the time to teach me and answer my questions made a huge difference. **Having diversity in the workplace is vital in many ways, including having a wide pool of mentors that young and new employees can identify with.**"

To continue her studies, Maya took out a student loan, emptied her savings, and went to the University of Dundee in Scotland to study remote sensing. At the time, many of her contemporaries were looking into graduate programs in Europe and Latin America, which were not as expensive as comparable programs in the United States and offered a high-quality education. Also, she says, she was drawn to the adventure of living in another country. When she returned to Puerto Rico with her master's degree, she landed a job as a cartographic technician in the GIS and Remote Sensing Lab of the institute, a position she held for more than 15 years.

During that time, Maya created hundreds of maps, but her proudest achievement is cocreating the *El Yunque National Forest Atlas* with colleague Isabel Pares—a massive endeavor for which she had to learn Adobe InDesign and upgrade her Illustrator skills. "We had virtually no budget to work with," Maya says, "so we did everything in-house—the writing, the cartography, the layout, the photos, the Spanish translation. It was a lot of hard work and long hours, and I am incredibly proud of what we produced." The atlas is a public document and freely available online in the USDA Forest Service publications repository, Treesearch.

Fun facts!

One thing that boosts her well-being: "My cats—they make me smile even on hard days."

Fun thing on her desk: "Cat-chewed Post-It love notes from my husband."

Maya, age 4, with her mother, Sylvia Zavala Trias, in San Juan, Puerto Rico, 1983.

Maya worked hard on helping create the *El Yunque National Forest Atlas*.

Credit: Gary Potts.

Building collaboration

In Maya's career so far, from student volunteer to program manager, successful collaboration has been key. "Collaborations have the added effect that you are working not just for a goal but also for your colleagues," she notes. "I don't like letting my team members down, and that motivates me to work hard and deliver something I am proud to present to my colleagues. It takes time to build relationships, and it's important to earn and keep that trust." Working successfully with others has special significance for Maya as a Latinx woman: "I am also motivated by my minority status. I often feel that if I don't work hard, it reflects badly on all my fellow Puerto Ricans. In my experience, we humans tend to generalize way too much, so I carry not just my reputation but also the reputation of whatever group others see me as belonging to—Puerto Rican, Hispanic, Latinx, Caribbean, woman … ."

Being pigeonholed is one challenge that Maya has had to confront in her working life. Another is her fear of public speaking, which, as an introvert, she says she has struggled with. But, she says,

"If you hide away from every opportunity to engage with others and showcase your work, you will never improve." The only way to overcome this fear, she believes, is practice: "Start with presenting to your friends. Also, facilitating meetings is a great way to improve your stage fright." Putting yourself out there may be uncomfortable, but Maya has learned that "what truly makes you better at public speaking is practice."

Maya photographing a caterpillar in El Yunque National Forest, 2019.

Credit: Gary Potts

"There is no better time for women to work in STEM fields. Great and brave women have paved the way for us to be able to work free of harassment and with recognition for our achievements. Be assertive, don't doubt your intelligence or ability, don't apologize for your gender or your culture, form and nurture relationships with other women at work, and protect each other. If you witness any type of harassment, always, always report it. It will not stop otherwise, and silence helps normalize unwanted behaviors."

On the other hand, Maya sees her greatest strengths as flexibility and the ability to collaborate with others. As the COVID-19 pandemic has reminded us, we live in a rapidly changing world, where we need to be able to adapt quickly to new working conditions and processes, she says. At the same time, **"we are working with shrinking budgets and larger scale problems, such as climate change. These big problems and our limited budgets require that we form alliances and work together to have an impact."**

With these challenges in mind, Maya has a hands-on definition of success. Success, to her, means "working towards something": "I have been in many meetings where a lot of great ideas get talked about, but without someone leading the charge and working long hours, those ideas do not go anywhere. They are just chatter in the wind." So, for Maya, success is "leading, working, creating" and having the flexibility to let your goals "move and morph" with new information and the changing times.

ALICE RATHJEN

Going on a spiritual journey to map genomes

Position

CEO and cofounder
DNA Compass

Education

GIS certificate
Columbia Junior College

BA in religious studies
UC Davis

*I*T WAS A TOMATO that changed everything for Alice Rathjen, CEO and cofounder of DNA Compass, a start-up company that uses GIS to map and manage genomes—a ground-breaking way to map human DNA. Originally intending to become a lawyer, Alice had been majoring in religious studies at the University of California, Davis, when researchers there turned off the rotting gene in a tomato, introducing Alice to genomics. **"I was obsessed with the creative potential of this new technology and its moral implications,"** the entrepreneur says.

Although she had an idea of what she wanted to do after being introduced to genomics, the path wasn't nearly as clear. For graduate school, she decided on Stanford to focus on genetic ethics—her plan was to become a professor. After a year and a half, though, Alice realized that no one was interested in that field at the time, the mid-1980s. **"It was clear to me that in the United States, the issues around genetic ethics would be sorted out in the marketplace, not by academics writing papers,"** she says. **"I resolved then that if I wanted to have an impact, I should eventually start a company in the genomics sector.** So I dropped out of Stanford to understand business and took a job in advertising. The advertising job was followed by a job in illustration and then a job illustrating site maps for an archaeology firm, which then led to studying GIS."

Discovering GIS

At Columbia Junior College, Alice had her first exposure to GIS. Living in a tiny cabin in the Sierra Nevada in California, she worked as a potter while completing her GIS certificate. "This was the early '90s. We had to create the map layers in AutoCAD, clean the drawings, and then create topology to get the maps to work," she says. "I fell in love with the technology and wanted to learn how to study cancer cluster analysis (in the hopes of preventing breast cancer). [But] we needed local/regional genomic datasets and consented genomes [cleared for use] so we could eliminate the genetic variable in the study of disease."

Alice kept studying genomics as a hobby in the hope of combining the two fields. Eventually, the quality of genetic data improved as did the capacity of mapping software. In the late 1990s, it occurred to her that mapping software could be used to map and manage genomes. Her religious studies training came in handy, because at the time, researchers thought genomes didn't change. "I thought the notion of a static genome was a projection of the soul into the cell. I knew anything alive changed and where things were inside the cell in relation to one another mattered. I further imagined genes functioned in the same way as watersheds, and that once they hit a density of mutations, they would flood the cell with proteins or cause the cell to go up in flames," Alice explains.

"I knew GIS could manage genomes down to the base pair level and that it was capable of implementing any changes in government regulations for who could access information. My hope was that location-based map layers would provide interoperability and support a multistakeholder environment."

Alice says success to her is "doing no harm and hopefully giving back as much or more than I've been given."

Alice painting outside with her son.

Facing hardship

In late 2000, Alice started Dominga, the first direct-to-consumer genetics testing service. Alice named the company in honor of the woman who provided her day care while she was growing up, Dominga Rodriguez Cortez. Although Dominga was early to market, it took six years for Alice's patent on mapping the genome in 2D and 3D to be issued. Competitors such as 23andMe, the Broad Institute of MIT and Harvard, and the Allen Institute raised massive amounts of funding and publicity, leaving Dominga outspent. "I didn't anticipate the market forces for selling deidentified patient data would be so strong," Alice says.

In 2009, Alice cofounded DNA Guide, which created the first genome browser for mobile platforms, with her friend, Saw Yu Wai. A big hurdle the company faced, though, was from academics patenting each incremental interpretation of the genome in the form of gene patents. "The gene patents prevented me from being able to add the annotation to the genome maps," she says. "Gene patents were thrown out in 2013 with annotation finally becoming publicly available. However, once gene patents were lifted, the FDA then started treating genome mapping software as if it was a medical device—requiring a difficult approval process for getting to market."

But the biggest hurdle came when Saw Yu died of pancreatic cancer in 2015. "She was brilliant and generous with her expertise and a pure joy to work with. She died three months after her diagnosis of pancreatic cancer. We tried to get her tumor data to analyze it using GIS for actionable markers, but it was too difficult to get the data and she deteriorated too fast. Losing her was like losing my right hand," Alice says. After that, she took a break from the project for a few years but came back even more motivated.

Saw Yu Wai, Alice's DNA Guide cofounder and friend.

Fun facts!

Fun thing on her desk: "I sometimes put the book *The Little Engine That Could* under my laptop when working on tough problems and put on my lucky shirt."

Favorite thing about GIS: "Maps combine art and visual intuition. But I wish map objects also had sound. It would allow us to incorporate information from another one of our senses."

A groundbreaking start-up

In 2018, along with Xavier Thomas and Bill Kimmerly, who have worked with Alice for more than 20 years, Alice launched DNA Compass, continuing the original effort of setting a global standard for informed consent enabling the use of human genomes in medical research. Considering that the academics' tools failed to enable patient engagement and local/regional control of genomic data, "We thought it was time to try again," she says. The company's technology is designed to protect patient privacy and build a "universal health record" for personalized medicine, Alice says. DNA Compass's mission is to build local and regional genomics capacity through the use of genetic data, using GIS to map and manage genomes within countries and communities around the world. The idea is to speed up health outcomes, particularly in situations such as COVID-19, in which data is limited. Alice wants to turn patient genetic information into a kind of location data so that researchers can visualize risk factors.

A dashboard from DNA Compass that distributes genomic variant information associated with COVID-19 and SARS-COV-2 virus exposure.

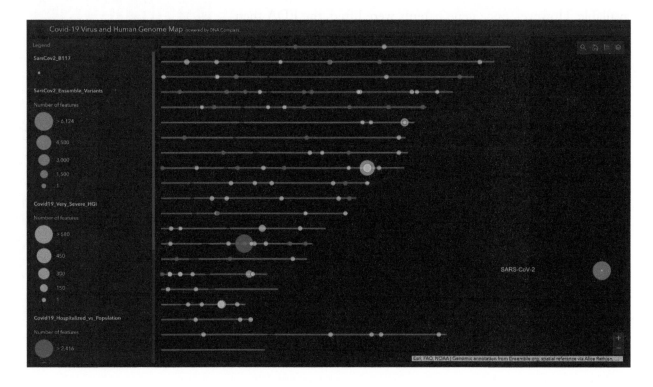

"Currently, genomic research, and in particular, COVID-19 research, is thwarted due to a lack of global genomics infrastructure that is capable of complying with regional government jurisdictions and patient consent," Alice says. **"I'm hoping population health professionals that currently successfully use ArcGIS Online for sharing health data across borders can help lead the way to make more genomes accessible to researchers."** For now, Alice has a pilot project in Nigeria, in partnership with the Christian Medical Scientists and Basic Health Foundation and Esther Moore of Sambus Geospatial, where they're working with COVID-19 genomic data in the hope of finding drug targets and clues for how to stratify the population of Nigeria on the basis of genomic risk. DNA Compass has also created a map (dnacovid-19.com) showing that it's possible to track specific mutations for the COVID-19 virus per person as well as what mutations a person might have in their genome that could guide treatment.

Creating the future

"I think it's the difficult people we encounter that make us driven, yet there are those who have been kind to us, who allow us to persevere," says Alice. She credits a long list of teachers and coaches who have helped her over the years, at one point even keeping a list in her wallet of more than 60 names. **"I grouped people into three categories: saints, survivors, and characters (the people who were authentically themselves),"** she says.

Among those who have inspired her is her mom, who died from breast cancer when Alice was 18. "My mom raised us that we had a moral obligation to not just vote, but also to participate in creating the future. Everyone had to 'go to bat' with their ideas. It didn't matter if you struck out. Time would sort the good ideas from the bad. My whole life I've 'gone to bat' and struck out many, many times. So … my personal motto is you go to bat enough

Alice working on a project with her son. She says she always tries to choose him over her work.

times … sooner or later, you'll get hit by the ball and end up on first base. Right? Anyway, that's what I keep telling myself."

As for what advice she'd give to young girls, Alice says, "I'd tell them to definitely learn GIS. **Humanity is like a flock of birds flying over a lake. Technologies like mapping software allow us to see our reflection and, hopefully, become more conscious and coordinated in our movements. We need more women in GIS to guide us through this next phase where we start making better decisions about our survival."**

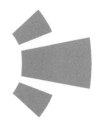

MARIA-ALICIA SERRANO

Bridging communities using insights from GIS

Position

Senior director of research, analytics, and insights
YMCA of the USA

Education

MA in public affairs
Indiana University, School of Public and Environmental Affairs

BA in law, letters, and society
University of Chicago, Illinois

*T*HE BREADTH AND DEPTH of the YMCA's mission gives Maria-Alicia Serrano the freedom to focus on a wide range of projects. As senior director of research, analytics, and insights at Y of the USA, she oversees insights and strategy for the development of timely and actionable data for the Y movement.

"My particular focus is increasing data accessibility and comprehension and insight delivery. I lead the production of reports, visualizations, geospatial tools, dashboards, and resources that support data-driven decision-making across all program and operational functions. Given the impact of the COVID-19 pandemic, my current focus is on supporting Ys in understanding their communities' current and projected needs," she says.

Maria-Alicia says there is so much to love about her job. The best part, she says, is being able to help people who haven't used data in their work before and, in some cases, almost have a fear of using it because they don't want to use it incorrectly. "When they realize that being a nonprofit leader who uses data doesn't mean abandoning their heart and passion for serving their community, it really moves me," she says. "In fact, the hardest part of the job is not being able to provide every single YMCA and community partners with every piece of data or insight tool that they could imagine."

Her greatest strength is the ability to synthesize vast amounts of information into digestible bites for those who

Maria-Alicia in Paris with her husband, Todd.

may not have subject matter expertise. "I know it is a cliché," she says, "but the ability to communicate with a variety of audiences, especially those who do not have the same level of knowledge of a subject as you, is key, no matter what your chosen career is."

A supportive start

In elementary school, Maria-Alicia wanted to be the president of the United States because she wanted to change the country. She feels blessed that she and her brother grew up with loving, supportive parents in a thriving community. "My parents believed that being a student was my job, so they did not want my brother and I to work during the school year, and even during the summer, we were told to take time off and relax," she says. "This gave me the luxury of being able to pursue various passions and interests throughout high school, college, and graduate school. They were adamant that we should pursue a career that we were passionate about."

Maria-Alicia's career is a combination of both of her parents' influence. Her father had his own real estate firm that served parts of the community that had been historically redlined, and her mom is a minister, social science researcher, and program evaluator.

Between the two of them, Maria-Alicia developed a passion for using qualitative and quantitative data to inform decisions on how to improve communities.

Maria-Alicia loved being in school and learning about new issues, topics, and cultures. Her broad range of interests was one of the reasons she selected the University of Chicago. While there, she narrowed her focus to policy development and implementation. Next, she earned a master's in public affairs from the School of Public and Environmental Affairs at Indiana University. While getting her master's degree, Maria-Alicia was influenced and supported by Orville Powell, her adviser, instructor, and mentor. "Orville is the ultimate example of a servant leader," she says. "It's not about his advancement but the advancement and growth of your staff and the communities you serve."

Maria-Alicia, *left*, with her mother, the Rev. Dr. Maryalice Newsome, and her father, Travis D. L. Newsome, at the ordination of her brother, the Rev. Travis A. Newsome, *second from left*.

Building bridges between communities

Maria-Alicia began her career conducting organizational strategic planning and reporting for the Chicago Housing Authority (CHA), the third-largest public housing authority in the United States. She was at the CHA at the height of its revitalization work.

"I loved the CHA, but the hardest part was the fact that our focus was on providing quality housing for individuals, which doesn't take into account the variety of other needs from education to childcare to access to grocery stores," she says. "After five years, I decided that I wanted to understand communities across the country as well as do work that looked at improving the resources available within and near communities. I joined Applied Real Estate Analysis (AREA) Inc., a boutique consulting firm based in Chicago."

While at AREA, she says she had her greatest professional growth as she learned "the delicate dance of conducting unbiased research for the people who paid you to conduct the research and how to conduct action-focused research." She was able to work with federal agencies and state and local agencies across the country.

Maria-Alicia says she is passionate about building bridges between communities, especially between White and Black people, by expanding awareness and understanding of systemic racism and its impact on individuals and generations. Although she would not describe it as a personal obstacle but an obstacle of society, she says, "I am a Black woman in a field with a larger number of men and White people. I walk into professional conferences or meetings and scan the room to often find I am the only Black woman; I've held telephone meetings with people, and the first time they see me in person, I get a look that makes it clear that I don't fit the mental image that they built. **Early in my career, I felt the need to help people feel comfortable with the fact that the researcher/consultant they were working with was Black. As I grew more confident, I quickly began deciding that it was not my job to make other people feel comfortable.** I use my voice and my position to ensure that the perspectives of Black women and men are taken into account."

As she hopes to lead by example, Maria-Alicia talks about the woman she admires most: her mom. "She is literally the smartest person I know and does research just for fun. She was in corporate America at a time when there were very few women in full-time professional positions, let alone Black women. Although she already had an MBA, she decided to get her PhD while I was in middle school. She leads by example in terms of pursuing your passion, even if it sends you on a very unique path that isn't in line with how the rest of the world thinks or operates. Her work lies at the intersection of public policy, ethnic and racial studies, and theology. She is the epitome of an optimist and unconditionally believes and supports me. That doesn't mean she won't call me out though if I am wrong: she's my mom—that's part of her job."

Maria-Alicia speaks at the Esri Federal GIS Conference in 2020.

In her career, the three accomplishments Maria-Alicia is most proud of are launching the first analytics and insights hub for the YMCA Network; authoring an action-focused research study on racial discrimination in several counties that included recommendations adopted by local governments, ultimately leading to policies and practices that helped reduce racial discrimination; and developing analytical models to predict socioeconomic conditions in several communities to help inform resources allocated by the Department of Defense to active military personnel.

But she also emphasizes: "Success to me is feeling joyful about my life. What's key to understand about that though is that **joy and happiness are not the same thing. You can be joyful no matter what the circumstances around you are, whereas happiness is dependent upon circumstances.** My faith in Christ is at the core of everything I do, whether it be personally or professionally. I want to be an example of Christ's unconditional love in all areas of my life. I want everyone with whom I interact to feel that I respect them and honor them as an individual."

Maria-Alicia's other motivators are her husband, Todd, who is her biggest advocate and has always encouraged and supported

Fun facts!

Hobbies: "My two hobbies are crocheting and photography. They allow me to be creative although I prefer photography because you get a more immediate return on your time as opposed to the weeks most crochet projects take."

What boosts her well-being: "Sitting on the floor with my husband and daughter coloring and talking about our day."

Career: For a hot minute, she wanted to be a fashion designer.

Maria-Alicia's daughter, Zariah, points to a sign created by members of their community, indicating their support for Black Lives Matter.

her career, and her daughter, Zariah, who motivates her to make the world better for the next generation. Maria-Alicia's advice for young women starting in the STEM and GIS fields: "Understanding the impact of place is key if you want to see changes in society. If you care about children growing to their full potential, GIS allows you to understand what resources a child has access to that may impact their long-term life outcomes. If you care about the physical and emotional health of marginalized populations, then GIS can help."

ALINA SHEMETOVA

Energizing GIS from a legacy of science

Position

GIS supervisor in advanced analytics
SM Energy Company

Education

Bachelor's in geography and environmental studies
University of Colorado, Denver

*I*N 1991, WHEN UKRAINE declared independence from the Soviet Union, Alina Shemetova was only an infant, but over the years her family told her stories of their hopes for the future. They lived in Sevastopol, Crimea, and they believed that freedom from the restrictive Soviet system would change their lives for the better. But five years later, the new Ukraine was mired in chaos and corruption with not enough work, resources, or stability. Frustrated by the lack of opportunities they had anticipated, Alina's parents decided in 1996 to move to Buenos Aires, Argentina. Their long-term goal was to make their way north to the United States.

Although Alina's father, Yuri Shemetov, was an established artist and her mother was well on her way to a career in electrical engineering, they chose to leave it all behind to take jobs as a contractor and a cook's assistant, respectively, determined, above all, to create a better life for their family. Alina, only six years old, remained under the care of her grandparents in Ukraine.

"The woman I admire the most would have to be my mother, Alla Shemetova," Alina says. "Starting her career over at 37 could not have been easy. Making the sacrifice of leaving me in Ukraine to start over in a completely new country and culture could not have been easy either. However, she would still find a way to become fluent in the language of each country she lived in as well as to support our family financially. No matter where life took her, my mother

Alina lived in Ukraine until she was seven years old. Here, she is four in her kindergarten picture. Sevastopol was a naval town, and young children often dressed up in sailor uniforms.

Alla Shemetova began her career as an engineer, though she had to reinvent herself in Buenos Aires and then again in Colorado as a paralegal and tax preparer when her family moved to the United States. Alina's father, Yuri, still creates art. The couple even got to achieve their dream when they had an art gallery for a time in the US.

was so optimistic about the prospect and would put everything she had into making any circumstance work. Watching her through the years has taught me to adapt to any situation and to do what you must in order to make an opportunity for yourself and your family."

Alina's grandpa, Dr. Valentin Bryantsev, became her biggest influence as she was growing up. As an oceanographer, he was part of the first joint US–USSR effort to research Atlantic fisheries, from 1967 to 1973. His visit to Woods Hole Oceanographic Institution in Massachusetts during that time gave the family an early glimpse into a new world of opportunity outside Ukraine.

Alina's aunt, Dr. Yuliya Bryantseva, also took young Alina under her wing. Over the span of her 30-plus years of research in the ecology of marine phytoplankton, she took eight oceanographic research cruises and completed work in the Black Sea, Mediterranean Sea, and Arabian Sea. Between visiting her grandpa's Azov and Black Sea Scientific Research Institute of Marine Fisheries and Oceanography in Kerch and her aunt's Institute of Biology of the Southern Seas in Sevastopol, Alina spent a majority of her childhood years staring at marine creatures through a microscope or thick panes of aquarium glass.

More than anything, Alina was inspired by the vigor and pride her grandpa and aunt showed in their work. She admired them for creating opportunities for themselves under a difficult Soviet system to continue their research. Their dedication to their professional work set an example for Alina that she has followed in her life and career.

By 1997, Alina's parents were settled into their new life in Argentina and brought seven-year-old Alina from Ukraine to live with them. Within a month, they got the call that they had won green cards in the US immigration lottery. They were now on the path to US citizenship and on May 23, 1997, moved to Leadville, Colorado, where they had friends.

Above: Alina at Balaklava Beach, near her Aunt Yuliya's research site in Sevastopol, Crimea.

Right: Alina stayed with her grandfather, Valentin, in Ukraine while her parents went to South America to start a new life.

Alina reflects on how she embodies the Ukrainian character: "Like most Ukrainians, I have a constant hunger to prove that I am capable, no matter the circumstances. It is a very cultural trait but not one that always flourished, especially within socialist/communist countries. Now I am in a country that allows you to be rewarded for it. I am always going to remember what it was like for my parents not to have had that and how much they sacrificed to provide me with a different path."

Clarifying her goals

Alina did not move away for college at the University of Colorado, Denver. She lived at home to save money, and since she had been working since 15, to stop working while in undergrad seemed unfathomable. She went to classes primarily two days a week and juggled two jobs outside of school as well as a position at the tutoring center in between classes. It seemed like every waking hour was filled with either work or school.

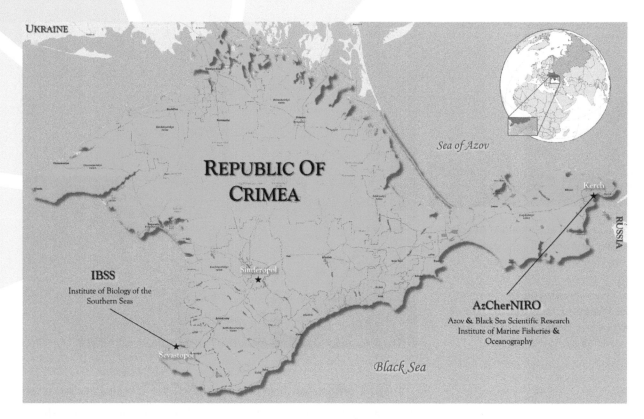

The places in Crimea where Alina spent her childhood with her grandpa and aunt pursuing scientific research.

In the summer of 2010, her aunt invited Alina to join her at the 45th European Marine Biology Symposium to see her present a scientific paper at Heriot-Watt University in Edinburgh, Scotland, when, Alina says, **"somehow everything I had learned and experienced up until then came into clear focus at exactly that point. I knew the passion I had for geography, technology, and the ability of the two combined to tell a story while driving substantial value."** When Alina returned to school, she chose her major in geography and environmental studies and enrolled in her first introduction to GIS course.

Undergrad was a time when, like most people, Alina spent a lot of time going in different directions. But two professors managed to grasp her attention and ultimately steer her on the path to GIS. Professor Deborah Thomas awed Alina with her research on hazards and health geography and her ability to use maps to communicate across cultures and disciplines. Deborah sponsored

Alina's internship at the Regional Transportation District (RTD), a project that was meant for only three months but, happily for Alina, evolved into a full year. Professor Rafael Moreno introduced Alina to natural resource management and the power of location-enabled data and technology. "Using geospatial technology has enabled me to translate the power of location intelligence into business value throughout my career at SM Energy," Alina says.

By the time Alina graduated from university in 2013, she had already paid off a substantial amount of her tuition as well as purchased her first home. Looking back, though, she wonders what it would have been like if she had immersed herself in all that the University of Denver had to offer. What it would have been like to learn from living in a dorm or closer to downtown Denver. Where her career would have taken her had she had more time to explore her interests such as oceanography and marine biology that had first consumed her while growing up in Ukraine.

A recruiter helped Alina set up an interview at SM Energy, an oil and gas exploration company. Alina was interested in the energy industry because it was at the intersection of resource management and geospatial technology. She was also aware that, for more than a century now, the availability of and access to natural resources to create energy have been intertwined with security and geopolitics, something that was becoming apparent in Crimea and would eventually drive her extended family to move north to Kiev, where they could hold on to their Ukrainian citizenship in the midst of Crimea's annexation by Russia in 2014. The interview for a land tech position at SM Energy was going well, and Alina was excited. She showed her maps, and the executives said, "Hey, I think we have a GIS department. ... Not sure what you're trying to show here...but you have the job!"

Visiting Aunt Yuliya, *left*, in Edinburgh was a life-changing trip when Alina suddenly realized her passion for geography and technology and using the two to tell stories that would help people.

Alina is a competitive tennis player. A good game helps her relax and let go.

Overcoming challenges to success

What Alina didn't realize at the time, and would come to know well, was the gap in the company's awareness of GIS and therefore its understanding of the value and practical applications of a location intelligence program. Translating the value of location intelligence to the business would become her mission and one of the biggest challenges she would face at SM Energy. But Alina likes a challenge, and keeping at it to achieve her mission, she would need to find the confidence to take a seat at the table.

"When I first started my career," she says, "even back when I interned at RTD, people would often tell me to slow down. I took this personally and was offended at those who I thought were trying to tamp down my flame and kill my motivation. **I was young and ambitious and often made mistakes. But thanks to those mistakes and the leaders who continued to guide me despite them, I was able to progress along the steep learning curve of applying my passion and education towards real-world problems.**"

It took almost seven years for SM to develop a GIS department. Before July 2019, Alina was responsible for driving the direction of her team and ensuring completion of tasks and stability in the services they provided—all without holding the title of manager or supervisor. Without the structure of a department in place to even warrant the need for a manager, Alina felt insecure about the value the company placed on her team and on her career development.

Letting this insecurity get to her made it hard to be firm with priorities and communicate confidence to her team. "I tried to accommodate everyone's needs and wants rather than focus on the skill set and actions required for the successful completion of a task," she says. "I was also much younger than the people that came to see me as their leader and the only female on the team at the time."

Yet Alina learned and continued to search for avenues to highlight her team's ability and the true value of having a dedicated GIS department to enable location intelligence for the company. For Alina, challenges bring opportunities. For the GIS team, this meant partnering with the business to enable location-based data collection and reporting on equipment used at oil wells. Capturing such data in real time reduced paperwork and increased awareness for the lease operator in the field and the engineer in the office. It also enhanced safety by tracking workers' exposure to H_2S (hydrogen sulfide), dangerous even in small concentrations. By implementing this type of collection and reporting, the GIS team helped SM achieve two strategic goals: creating a safer workplace and saving upward of 30 percent on the $10 million chemical program.

Another big win for the GIS team came from the implementation of yet another location-based field app used to capture data on sand flowback at each well—saving the company an estimated $40,000 per new well. With these accomplishments, the GIS department was established and valued companywide, and Alina became supervisor.

Alina attributes her success to her stubbornness and grit coupled with her impatience for improvement. Plus, she has grown

Alina speaks about location intelligence with return on investment at the Petroleum Users Group conference in Houston in 2019.

as a leader and a person. "I have grown to thoroughly enjoy criticism," she says. "I will take advice and feedback from anyone willing to go out of their way to give it. Good or bad—does not matter. I understand that those who present criticism or advice may be as vulnerable as I am in that situation so I try to make a concerted effort to show that I am not on the defensive and communicate that there is always something to learn from every situation." Practicing delayed gratification also helps, she says, as success is like building one brick at a time. "These traits have paved the way to my success—or what I like to think of as doors opening to more opportunities to participate in new experiences that contribute to my lifelong learning," she says. In early 2021, she was awarded a GIS Professional (GISP) certification after studying for more than four years and passing the exam in December 2020.

Alina encourages all young people to consider studying more geography. "Geography is an inclusive science that is both a discipline and a tool used to deal with a variety of real-world issues that affect people," she says. "It can bring people together for collaboration and can allow creativity to flourish. It is not set in a context of 'this is how it's always been,' so people tend to gravitate towards it. Also, data is fun, and the technology we have to explore it is evolving at a precipitous pace, but the spatial component really helps bring that data to life."

ARIELLE SIMMONS-STEFFEN

Protecting watersheds for generations to come

S SHE WAS GROWING UP, Arielle Simmons-Steffen's goals were modest if not vague. Mainly, she says, the plan was to "get out of Dodge and live in a big city. Maybe as an actor or a writer." Looking back, she realizes she did not really know anything about the types of jobs that existed in urban settings, having lived her entire life in Norco, a small rural town nicknamed "Horse Town, USA" in the Inland Empire of Southern California. Beyond accounting and teaching, most white-collar careers were a mystery to her as she grew up. Still, living on a ranch where water use and droughts were a constant concern fostered her future interest in resource planning.

Today, Arielle works at Seattle Public Utilities (SPU) to ensure that everything is being done to protect and maintain the environmental health of Seattle's water infrastructure. As an environmental planner by trade and training, she does not consider herself a "GIS person" but a data-driven planner. She prides herself on being able to break down information silos and increase data accessibility.

Some of her recent major projects include analyzing changes in public water usage during the COVID-19 crisis, creating baseline evaluation criteria for upcoming floodplain projects, and creating site protection criteria for SPU-owned properties. Arielle prefers working in the field rather than in

Position

Civil engineering specialist
Seattle Public Utilities

Education

MLA, landscape architecture and environmental planning
University of California (UC), Berkeley

BAs in geography and English
UC Berkeley

the office, she says, but in 2020 she spent most of her time working from her home office, which she shares with her husband and two children.

In 2020, Arielle also developed treemama.org and created and launched a tree mobile mapping app called iSeaTree. She founded these projects on the belief that kids can do more for environmental science than just learn about it—they can participate, too. Her previous experience as both an educator and an open-source software developer helped her guide both projects into becoming useful community resources during the COVID-19 era of remote learning. She is also involved in hands-on curriculum development, technology workshops, community presentations, and Code for America and the DemocracyLab hackathon project management.

Arielle enjoys working outdoors.

Small-town hometown

Raised by a single mom, first in her grandparents' home and then at the ranch her mother purchased in the 1990s, Arielle was shaped and inspired by her childhood experiences. Even though her mother, Jennifer Simmons, was busy with a full-time job and

Arielle is passionate about building a world where all kids, including girls, can analyze their natural environment and actively participate in its protection. She developed treemama.org, see logo *above*, and created the iSeaTree app, on phone at *left*, during the COVID-19 pandemic to involve children in environmental science.

Arielle flags vegetation for Seattle Public Utilities on a floodplain reconnection project.

college classes, she would still find time to go horseback riding with Arielle. Their favorite ride was always to the local riverbed, to the Santa Ana River, which, years later, would become the inspiration for Arielle's master's thesis.

Along with her love of the outdoors, Arielle was exposed to computers early on through her grandfather, Kenneth J. Simmons. Kenneth was always tinkering with computers and technology. Arielle recalls being one of the few elementary students in her hometown who had a computer in her house during the 1980s. Arielle's

Like her grandmother, *left*, and mother, *right*, Arielle (on horseback) is a huge horseback riding fan.

Arielle's grandpa, Kenneth J. Simmons, exposed her to computers at an early age.

grandfather was also familiar with GIS techniques, and in his later career became a systems engineer at Lockheed-Martin working with GPS data and uncrewed aircraft technologies.

Almost every member of her family was a math whiz. Arielle, on the other hand, had more of a natural affinity with reading and writing. But, she says, "sometime around college, math became more of a core interest when I started working with research data. Once I mastered how to tell stories with scientific data, I was finally hooked on math. One of the reasons I started treemama.org and created the iSeaTree app is so that young K–6 students, like my daughters, have a way to discover math through data literacy."

Words of wisdom

After attending a community college, Arielle applied to three state schools. She got into her second choice, UC Berkeley, where she planned to study geography and English. But going to a massive research university 450 miles from home was a leap into the unknown. Most of her college-going family members had remained near home and gone through smaller and more trade-specific programs. Also, she says, her family tended to ridicule academic researchers, mostly because they studied obscure topics and were rumored to have low salaries. Her stepfather, she says, never saw much value in academia and had a habit of taking away Arielle's schoolbooks when he felt she was focusing on them at the expense of her ranch chores. Also at UC Berkeley, Arielle discovered that she would be one of the few in her academic cohort who had to work part time at a minimum-wage job while taking on a full-time course load.

Before she left for college, Arielle confessed her worries to her grandfather. He was a high school valedictorian who had overcome his own family differences. Arielle says, "He sternly reminded me that my own father, a poor immigrant from Pakistan, had made a much bigger leap than I would be when pursuing his own education. My grandfather told me, 'Some people let the world tell them who they are. And then there are the people who are wise enough not to listen—they are the people who change the world.' I have valued this wisdom throughout my life."

Today, Arielle is grateful to both sides of her family—her grandfather, for showing her the possibilities in being academically accomplished, and her mother and grandmother, for showing her how to be resilient in the face of adversity. "Ultimately," she says, "it was my family who gave me the guts and fortitude to pursue the path so different from the paths they took and understood."

Support systems

At UC Berkeley, several teachers helped Arielle bridge the transition from ranch kid to academic, including urban forestry expert Joe McBride, former US Poet Laureate Robert Hass, water geomorphologist Matt Kondolf, and GIS expert Maggi Kelly. Professors Kondolf and McBride were both on her thesis committee. "They taught me important skills," Arielle says, "particularly how to take big ideas and translate them into workable action items. I was lucky to work with both—especially since both were good at promoting diversity within their own work groups.

Arielle, age seven, at a lake in the Santa Ana River watershed, an important landmark of her childhood.

Professor Kondolf employed many female graduate students for rigorous fieldwork assignments, something that was not always a guarantee in other graduate programs I looked at."

Equally important were the female friends she made in graduate school and in the UC Berkeley co-op system. They all supported her different background and were powerful role models, she says. Many of them, like Arielle, were balancing multiple roles within their careers and identities. "They exemplified leadership, multidisciplinary thinking, and always embraced the full potential of their life interests," Arielle says. At UC Berkeley, "all of these people—all of them with different approaches—found ways to help me become the person I am today."

Returning to the river

After Arielle left her dusty hometown for college, the Santa Ana River became the symbol of all the things she could not forget. During the late '90s and beyond, the struggle of urban and rural watershed management was rapidly changing her hometown. Water issues had always been a factor in her childhood, but now they were at the heart of a transformative exodus changing her hometown from being a primarily agricultural landscape. "As the dairies and chicken farms became replaced with tract homes," she says, "the riverbed, a place that had forged an early relationship between myself and my mother, became the last landmark of what remained of my childhood."

To understand what made rivers such powerful, unforgettable, and essential resources, Arielle spent her college years poring over maps of the Santa Ana River and mapping other California river systems. Her graduate thesis introduced the latest remote sensing techniques for mapping and managing the riparian species *Arundo donax*, a tall perennial cane, one of several invasive species that worsened flood damage and eradicated native species. To some, focusing her master's thesis on a far-off rural landscape problem such as *Arundo* might have seemed an odd choice, but for Arielle, it was the only choice. *Arundo* was what had created the unforgettable Santa Ana tunnels that she and her mother used to ride their horses through. It was also what had wiped out their local bridges during the El Niño storm years and decimated the habitat of their native fisheries.

Arielle's agriculture heritage gave her a profound respect for all that rivers do to shape and maintain natural environments. Though some people rightfully point out that agriculture is one of the biggest degraders to river health, Arielle sides with Chris Johns, former editor in chief of *National Geographic* magazine, in noting that agriculture is one of the few industries actively raising and educating our future water conservationists.

River systems and how they interact with residential populations, as both an environmental and cultural resource, have long been at the heart of Arielle's life's work. As in Arielle's family, the impacts of rivers travel across generations. Rivers change along with the land and alongside the people who live on that land, she says. **"Rivers are an apt reflection of who we are becoming as a species. They remind us of the cause and effect of our past decisions. And they are always striving to build an equilibrium between themselves and the lands they occupy."**

Start-up culture

Arielle came out of school as an environmental planner with a data background. Since then, she has experienced a myriad of roles ranging from private-sector software developer to forestry and water resource consultant.

One notable career experience was at a software start-up that was heavy with "brogrammers" who kept their beer in the fridge and maintained exclusive cliques that didn't include her. Start-up culture is known for being difficult and isolating for women, Arielle says, and she experienced that firsthand as the sole female software developer on an all-male team for three years. Compounding the problem, her entire management chain was male, and all the direct supervisors she had during her time there were also male.

At first, the position had a lot of pluses and offered interesting work assignments. As time went on, though, Arielle says she felt increasingly limited, disrespected, and excluded. In her third year, her direct supervisor took credit for her ideas without acknowledging her because, she recalls him saying, "he was new and needed to distinguish himself." Yet she shrugged off such incidents in the interests of fitting in with the team and having a harmonious work life—until her third year, when she was experiencing a difficult pregnancy. When she told her team she needed to take bed rest, she was publicly berated by a senior project manager who yelled, "Can't you work while on bed rest?" After her delivery, human resources repeatedly denied her and the few other new mothers at the company the right to an exclusive mothers' pumping room for privacy.

Many of the issues Arielle faced during her pregnancy and first year of motherhood opened her eyes to why there were so few women and mothers at that job and at many other tech jobs. "The void of women in my direct management chain had created a system where the interests of women, particularly senior women with children and who had some experience, had no framework of support," she says. **"It seemed to me that the system had been built to allow in entry-level women but not designed to keep them."**

After the birth of her daughter, regaining her professional place and voice in such an environment became increasingly difficult for Arielle. She knew that the office culture was limiting her and decided to move on.

Tip!

"No matter what field you pursue, STEM or not, the need to be an analytical storyteller is a universal skill. Learning how to distill a problem to the point of visualizing a result is the most valuable ability you can acquire in life."

Using all her skills

After her disappointing experience at the start-up, Arielle decided to return to public-sector employment, seeking a place where she could use all her skills, including the leadership and collaboration skills that she had gained as a mother. At SPU, she was relieved to find a culture where mothers not only worked but advanced as well. It was a welcome relief to go through her second pregnancy in a more respectful workplace than her first.

Arielle describes herself as a survivor. She says, "My mother and grandmother are both stubborn, resourceful, and wildly independent. They are both good storytellers, particularly my grandmother. Having been raised predominantly by these two women, I did not fall far from the tree. Our independence means we don't need much other than ourselves to succeed." Though independence is her greatest strength, she recognizes that the world today is facing difficult, long-term environmental problems that will outlast her individual career timeline. Accordingly, she is shifting her focus from independent to interdependent work, building team momentum, and designing transparent work processes that have longevity to work for long-range planning scenarios.

For Arielle, GIS is more than a technical skill set; it is a way of creating a framework for understanding problems in the physical world. She reminds women not to be afraid of technology and not to be dazzled by the newest, trendiest tools. Building a career that focuses on strengthening skills in collaboration, strategy, and critical thinking, she says, is what is important. **"For me," Arielle says, "GIS is more of a thinking process than a technology. Technology comes and goes, but the process of building a story, analyzing it, and sharing how it will end is really the heart and soul of what GIS is about."**

LAUREN SINCLAIR

Empowering kids using GIS

Position

Middle school teacher
French American International
School, Portland, Oregon

Education

MS in geography
Portland State University, Oregon

BS in education
Samford University,
Birmingham, Alabama

"*I* BELIEVE IN FUN. If it isn't fun, why do it?" says middle school teacher Lauren Sinclair. "I can make just about anything fun, from GIS to the 'suffer-fest' hikes I treat my nephews to. That's a value I try to teach my students: if something isn't fun, you can tweak it and make it fun." Sinclair, who teaches math, science, and technology for grades 6–8, embodies a passion for learning and fun. The students in her classes don't just map, they do hands-on cartography projects, using materials such as Play-Doh, cardboard, and calligraphy pens. She encourages students to explore GIS, its impact, and how to use it to create a map from scratch. **"If you are smart, driven, and want to get your message out into the world, chances are you should be telling part of your story with a GIS—a powerful, interactive map that opens people's eyes to what you have to say,"** she says.

"I want students to see that GIS technology can make a real impact on the world around them. When my students reach my eighth-grade GIS Analysis class we focus exclusively on ways their maps can make a difference. We start by studying gerrymandering and redistricting. Then the kids get a chance to take action on what they've learned. These days we're also studying how maps help us understand and stop the spread of epidemics. Every semester we participate in a Mapathon (from the Missing Maps project) to help communities in the wake of disease, natural disasters, and

Lauren hiking with her nephew in Zion National Park.

other humanitarian crises. My eighth graders end their semester by creating [an ArcGIS StoryMaps] story about any cause they are passionate about," Lauren explains. **"I truly believe that the point of middle school and high school education is to talk about what matters, then do something about it. I want my students to make the world a better place."**

Growing up imaginative

Lauren in her Where's Waldo? costume.

As a homeschool student, Lauren was able to spend a lot of time on creative pursuits, which translates to her teaching style. "If I was obsessed with old maps (I was!), I could spend a whole day exploring every detail of them, designing my own, then leading my brothers on a scavenger hunt based on the map I made. It was heaven," she says. "I try to bring that unlimited joy of discovery to my classes. Sometimes we stop projects just to play Where in the World Is Carmen Sandiego? or dig through Where's Waldo? books. I want to cultivate the pure love of learning that is so accessible in mapmaking."

Fascinated by the idea of foreign lands, exploration, and discovery as a kid, Lauren loved studying geography and history. "Since this was all before the internet really made it into our homes, I was doing all of this exploration through books or altases, so maps really unlocked each world I wanted to 'travel' to in my imagination. I kept the *National Geographic* maps that used to come in each magazine in a box under my bed," she explains. Although Lauren says she didn't personally know many women who were scientists, mathematicians, or experts in technology, homeschooling gave her confidence that she could find her own way through anything and become anything she wanted to be. "My most constant teacher, Mama, did everything and did it well … so I never thought I couldn't do something. I just figured I'd have to give my dreams a try," she says.

Lauren also credits her grandmother with her love of learning and geography. "From my perspective as a young girl, she was the woman in my family who represented a different path: she was an academic who pushed herself to reach her intellectual potential, earning a master's in library science and continuing into doctorate studies; she was a researcher and a writer; she started her own business," Lauren says. Always armed with a briefcase filled with well-creased maps, pencils, and yellow legal pads, her grandmother would have Lauren practice locating where they were, where they

Lauren with her grandmother, Ellen Miller, who inspired Lauren's love of learning.

were headed, and how she could measure nearby landmarks and directions. "When I chose to pursue a master of science in geomorphology, I used Grandma as my beacon, the only woman I knew who had struck off for grad school on her own," Lauren says. "That master's program led me to take GIS classes, where I developed the ideas I've now shaped into GIS courses for my students."

Becoming a teacher

Despite having never been to public school, Lauren says she always knew she wanted to be a teacher. "I think I must've imagined what school was like from TV programs," she says. For college, she chose a small private university in Birmingham, Alabama, that was well-known for its teacher education. "I adjusted to college really well because you have a lot of time to manage between classes, and that's what homeschooling is like. It was a fun time and an easy adjustment, though I challenged myself by taking extra classes and working really hard to get straight A's," she says.

She spent most of her time in student teaching placements in public schools in the area. When she wasn't working, though, she filled her schedule with geography classes taught by Dr. Eric Fournier, who sparked her imagination and inspired her to want to teach older students so she could teach social studies. Fournier also introduced Lauren to GIS, forever changing her teaching path. During a GIS activity on the Ring of Fire, a region known for its high volcanic and seismic activity, in her freshman geography class, Lauren was hooked.

"I immediately thought, 'Kids have to do this!' Back then, that seemed impossible. The first school I taught at in Birmingham, Alabama, didn't even have computerized projectors in each classroom," she says. "I had to write on an old overhead projector with wet-erase pens, then crank a wheel to move down the clear plastic 'scroll.' So it took a while for time and technology to catch up before I had the opportunity to use GIS in a classroom."

Lauren loves outdoor sports and spends her time rock climbing, surfing, and snowboarding, among other activities.

Fun fact!

Superpower she wishes she had: "Flight. I dream of flying from time to time. ... It's usually like calmly swimming breaststroke, and I usually end up in a tree. Maybe part of me wants to be a bird."

After a few years of teaching, Lauren moved to Oregon to attend Portland State University for her master's degree. When she was trying to decide where to apply for grad school, her mentor, Fournier, advised her to apply for every teaching assistantship she could find. She did, turning down any offer without a financial package. "Though I was nervous about it at the time, I picked the university that offered me a full ride, plus a housing stipend. That is the single best financial decision I've ever made in my career. It led to me being debt-free at a much earlier age than many master's program graduates I know. I don't know if that advice would still work today, more than a decade later, but I would advise any woman (we make less than our male counterparts), and especially any teacher (you probably won't earn a living wage), to follow this rule if at all possible," she says.

Bringing GIS to the classroom

After grad school, Lauren went back to teaching, still unable to incorporate GIS in the classroom because of a lack of technology—it required a powerful desktop computer at the time. She also worked as a national park ranger and state park ranger during her summer breaks, a childhood dream of hers. It wasn't until 2016, when she was asked to teach a technology design class at her current school, that Lauren was able to incorporate GIS. But finding activities and curriculum for kids proved difficult, and she had no idea how far GIS technology had evolved since 2010.

The big breakthrough came during Lauren's first Esri User Conference. Feeling completely isolated in what she was doing—nobody she knew was teaching GIS in K–12 education, let alone grades 6–8—she planned the boldest outfit she could find (a canary-yellow dress) and asked one question at the end of the education summit plenary: "Is anyone out there teaching GIS to kids younger than college? I'm a middle school teacher, and I need help." People swarmed her, pressing business cards into her hand and

"GIS can be whatever you want it to be. Want to be competitive in your dream job? I guarantee you that any job you want can be done better with GIS, and your ability to use GIS will make you stand out in a crowd. Want to make elections fairer in the United States? There are women who use GIS to fight gerrymandering and make sure that every vote counts. Want to design beautiful maps that take your breath away? There are women who use GIS to do that. Want to code and make lots of money? GIS can take you there. Want to make maps without coding? You can do that, too! Want to teach GIS to teens and tweens through fun projects? I hope so—we need more GIS teachers in the world. If my middle school students can use GIS in all these ways after just one semester, you can, too. Just give it a try and determine the direction you want to go."

Lauren in her bright-yellow dress at her first Esri User Conference.

spouting off names of people she should meet who had been working on the same idea for decades—that kids need access to GIS. "Over the next week, I got to know the small band of passionate GIS educators who I now call my mentors, colleagues, and friends," she says. "My conversations with them have transformed my GIS classes into a rich progression from learning 'What is GIS?' at age 10 to analyzing voter data and epidemics by age 14."

Now Lauren is an Esri educator consultant, creating educational content for K–12 students. "The day I decided to put on that yellow dress and stand in front of a crowd of hundreds of people was a big moment for me. I was summoning the strength I'd seen in my grandma, in teachers I'd taught beside, and in other women I'd met during my graduate studies. I'd never done something that bold, but the reward was incredible, so I haven't stopped hustling since." She's gone on to become a National Geographic certified educator, thanks to that boost of confidence and encouragement from the mentors she met at Esri UC.

"I have learned a ton and feel that my own ideas are amplified so other teachers can find them and use my activities in their

Lauren in her Carmen Sandiego cosplay during the National Geographic Educators' Summit in 2019.

Lauren takes inspiration from fictional teachers, including Ms. Frizzle, Anne of Green Gables, and even Bill Nye, the Science Guy.

classrooms," she says. And when she's not teaching her own classes, Lauren works with teachers in every subject area of her school to bring GIS into their classrooms. **"My dream is to expand my GIS education program so that every one of the 210 students at our school uses GIS every year in a variety of classes and settings,"** Lauren says.

She's proactive in her approach, keeping a calendar of "geo-inquiries" that match what each grade is studying in each subject at any given time. "That way I can email teachers and say, 'I see you're studying the Cradles of Civilization. I have a great interactive GIS activity for that! Would you like me to teach it while you watch? Then if you like it, you can do it yourself in the future,'" Lauren explains. And although she hates working on tiny computer screens, she's found the ArcGIS Online web-based platform to be a great selling point to teachers across the country who might be hesitant to try GIS activities at their school. **"When I show them that all they need is an internet connection and the ability to read carefully/follow directions, they're much more willing to give GIS a try,"** she says. An added benefit: because her classes are based on using ArcGIS Online, none of her students had any trouble accessing GIS activities at home when COVID-19 hit.

When Lauren first had to shelter in place in March 2020, she had to transition her GIS classes to a remote learning setting. As she was rethinking her curriculum, her eighth-grade students started asking her about quarantine and the pandemic; they especially wanted to learn more about the COVID-19 dashboards that were becoming popular. Lauren scrapped her previous plans and started teaching about the history of using maps to study epidemics. "After a few months, we had studied the Black Plague, cholera, and COVID," Lauren says, "and my students had created Story-Maps [stories] about what they had learned. They were enthusiastic that this was the best thing they'd done during quarantine." Wanting to help others study mapping epidemics, Lauren took those lessons and turned them into a video series with Esri's help.

According to Lauren, working with her "hilarious, creative, thoughtful, world-changing students" is the best part of being a teacher. "GIS is a powerful tool, and when it's placed in the hands of caring citizens, it can be used to create a brighter future. My students really care about the world's problems, and they love making an impact using real data in a GIS. It's easy to stay motivated when you're making a difference."

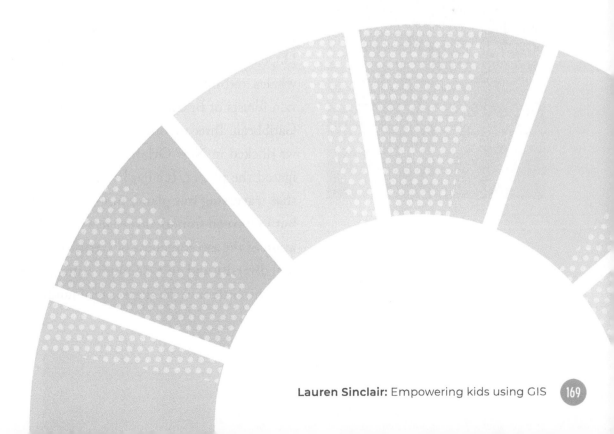

REGAN SMYTH

Seeing the big picture and keeping it real

HEN REGAN SMYTH was an undergraduate at Duke University in Durham, North Carolina, she participated in a study abroad program, Semester at Sea, that changed the way she saw the world:

We departed Vancouver on a steam ship carrying 600 students and faculty; passed the Aleutian Islands on the gray Pacific on the way to Japan; steamed on up the Yangtze to Shanghai in smog so dense we could hardly see our hands; looped through southeast Asia with stops in Malaysia, Vietnam, and India; and stopped to explore the plains of the Serengeti before traversing the rough waters of the Cape of Good Hope. We crossed the Atlantic accompanied by right whales and flying fish, visited the vanishing coastal rain forests of Brazil, then headed home through the Caribbean. Three months after leaving the West Coast, we docked in New Orleans, having fully circumnavigated the planet. **It's hard to articulate the impact that traversing our planet in its entirety had on me, but I returned from that voyage with both an appreciation of the awesomeness of Earth's human and natural diversity and an even more acute understanding that this little rock we share is profoundly finite.**

Regan directs NatureServe's spatial analysis program.

At that point, Regan was already well into a biology major, but inspired by her semester abroad, she added a second major, in environmental science. "I wanted to prepare myself to be part of building a more sustainable future," she says. Then she took an intensive class in GIS: "Twenty years later, I can still recall sitting at a computer in the college lab, watching the information come together on my screen, and thinking, 'This is so cool!' Ever since then, I've made sure GIS was part of my toolkit."

Charting her own path

In her current position as director of NatureServe's spatial analysis program, Regan leads a team of scientists and GIS analysts working to provide the scientific information that others need to effectively conserve biodiversity across the Americas. "We use spatial analysis tools, including GIS and R, to generate information about where species and ecosystems occur and how they are faring in a time of rapid global change," she says. "An important component of my work is providing scientific and technical leadership and team coordination for NatureServe's habitat-modeling initiative," a major scientific collaboration that resulted in publication of the Map of Biodiversity Importance. "Together, we are transforming the effectiveness of species conservation by generating more precise information on where at-risk species are likely to be found and using that information to guide conservation investment."

Looking back on her early education, Regan warmly recalls her high school English teacher, Sister Cherry, "who held us all to incredibly high standards in much the way you might imagine a Catholic school–teaching nun might." But when Regan was applying to colleges, Sister Cherry cautioned her that the sciences might not be the best fit for her talents. Regan admits, "I am certainly happier speaking to a crowd than working quietly in a lab. However, there are many different paths STEM fields can lead down, and

Regan gives a biodiversity presentation for NatureServe.

luckily, I did not let that advice dissuade me. I think Sister Cherry would agree that where I've landed is a good fit for my skills. My job today is as much about communication as it is about science, as I'm tasked with telling the story about what we do and why it's important, through both language and data visualizations."

Another formative experience during Regan's undergraduate years was her involvement in a program called Project WILD (Wilderness Initiative for Learning at Duke), where she participated in, and then helped lead, an experiential learning curriculum that included backpacking trips in North Carolina's Pisgah National Forest and a semester-long student-led course that explored leadership, risk-taking, and environmental stewardship. "My experiences in the wilderness as part of that program undoubtably contributed to my interest in pursuing a career in environmental science," she says, "but it was the study and practical application of leadership skills that I think had the greatest influence on me. **Project WILD taught me to embrace my own unique leadership style, identify real versus perceived risks, and push myself into new and sometimes scary experiences with confidence.**" In college, those scary experiences involved scaling rock walls and leading crews of inexperienced backpackers. Later, in her career, they have included pushing herself to speak confidently of problems and new ideas to senior colleagues and taking on leadership of ambitious projects that stretched her management skills.

Finding her focus

After graduating from Duke, Regan spent two years at an environmental consulting firm, performing ecological and human health risk assessments at contaminated waste sites. "It was in this position that I first started using GIS in a professional context, and the work was fairly interesting," she says. "However, the job made me realize that to be professionally happy, I should pursue work where I could completely focus on service to the common good instead of service to private industry. I also came away with an understanding that to make the most impact, it was necessary to think at the landscape level instead of site scale. When I left that position to return to graduate school, I looked specifically for programs that would enable a landscape ecology focus."

She found such a program at Duke's Nicholas School of the Environment, where, while completing a master's in engineering management in ecosystem science and management, she landed internships at two environmental nonprofits. She spent one summer working for The Nature Conservancy in Lander, Wyoming, and another summer mapping ungulate migration corridors in the Greater Yellowstone Area for the Wildlife Conservation Society.

That project, using GIS to analyze threats along those corridors, became her master's thesis. **"Those were incredible jobs for a 25-year-old with a love of the outdoors, commitment to conservation, and quest for adventure,"** Regan says. "They left me confident I had chosen the right career path."

Balancing act

After graduate school, Regan faced an all too familiar conflict. As she puts it, "Life called me back to the East Coast." It was a tough call: "My decision to leave behind opportunities to work in the wild western landscapes that captivated me was driven primarily by the fact that I met someone I wanted to start a family with, and his job kept him on the East Coast. I very much struggled with that at the time. **After a lifetime of being told 'You can be whatever you want to be,' I came face to face with the reality that, in fact, people often must make the choice between family and other goals. And let's face it, 'people' most often means 'women.'"**

At the time, Regan had a position as a regional ecological and GIS analyst at NatureServe's office in Durham. But when she started her family, having two sons within two years, she cut back to part-time work.

"For a period of about seven years when my children were young, I worked part time in a relatively low-stress technical position," she says. "I really enjoyed that time with my children, and my part-time schedule kept me and my husband sane. However, I also felt frustrated that my career had somewhat stagnated. I was very fortunate that NatureServe allowed me the flexibility to work part time and largely from home, but I also felt powerless to pursue other opportunities because I knew it would be a challenge to find that flexibility elsewhere."

Regan credits NatureServe's chief scientist, Healy Hamilton (also featured in this book), with encouraging her to return to

Regan and her sons, Caleb, *left*, and Loki.

full-time status when her younger son entered elementary school. And as her workload and responsibility have increased over time, she has found great fulfillment in her professional achievements. Even so, she says, balancing family and career remains an ongoing challenge.

"The best advice I can give to other women, and men, who may be in a similar position as they start their careers and family is to keep in mind that this is a long game," she says. "It is okay to not be making stellar professional achievements for a few years as you devote more attention to your family, and it is also okay—and even beneficial—for your children to have your attention at times pulled away from them and directed at other important matters. If family and career success is something you want for yourself, remember that you do not need to achieve all your goals at once— and also, choose a partner wisely."

Of course, Regan points out, "it is important to note the role that policies and practices play in supporting professionals with families. NatureServe's liberal policies regarding working from home and working part time made it possible for me to stay engaged, respected, and in a position to dial things back up when my other commitments allowed. Too often, as we are seeing now with the loss of women from the workforce during the pandemic, those policies are not in place, women are faced with impossible choices, and a great deal of talent is lost."

Regan also emphasizes that, in a technical field such as GIS, "it helps tremendously to find a way to keep your foot in the door if you are going to take some time off from your career. While the work I did when my children were young wasn't the most personally inspiring, it positioned me well to step up my game when I was ready."

Regan, *left*, and NatureServe Chief Scientist Healy Hamilton (also featured in this book), backstage at the Esri User Conference 2019 before the plenary.

Synthesizing skills

After transitioning back to full-time work, Regan took on a leadership role for NatureServe's network species habitat modeling initiative and moved into the newly created position of program manager for spatial analysis at NatureServe. Her greatest professional achievement to date, she says, has been her leadership role in the production of the Map of Biodiversity Importance.

"This project grew out of NatureServe's habitat modeling initiative, and Healy's vision, as shared with Jack Dangermond, that with an investment of resources, we could provide spatial data that could transform the conservation landscape. With investments from Esri, The Nature Conservancy, and Microsoft's AI for Earth program, we set out to model habitat for 2,216 of the nation's most imperiled species. The effort required bringing together the data and expertise of NatureServe's network of over 1,000 conservation professionals with machine learning and online collaboration technology. The outcome was new, precise information on the areas of greatest conservation need in the conterminous United States. My role in the project was far-reaching and grew as I demonstrated my ability to successfully navigate a large, complex, and highly visible project involving dozens of contributing scientists."

Tip!

"There is still so much to be discovered about this amazing world, and equipping yourself well with technical skills opens the door for you to be part of that discovery."

Regan at the 2019 Esri User Conference in San Diego, California.

Fun fact!

Fun thing on her desk: "There are several photographs on my office walls of amphibians atop mushrooms. On a backpacking trip in the Adirondacks, my sons and I came across a beautiful, fiery-red eft (newt) in the duff of the forest floor, close to a gnarly-looking shelf fungus. We positioned the little guy on the mushroom, snapped some photographs, and a tradition was born. Now on hikes, fishing trips, and walks in the woods, we search for different amphibians and interesting-looking fungi to photograph them with. To me, those pictures represent the amazing diversity of life that inspires my work (cool frogs and salamanders, weird spongy mushrooms). Because they are something I share with my children, they are also a reminder of the imperative to sustain that amazing diversity of life for future generations."

As director of spatial analysis at NatureServe, a position created specifically for her, Regan believes she has stepped into a role that maximizes her greatest strength. **"I am good at synthesizing information and experiences from multiple sources to see the big picture. In a workplace full of technical experts who are incredibly talented but sometimes narrowly focused on the task directly at hand, that skill has allowed me to contribute significant value.** Being able to step back to understand what information is going to be most valuable to an end user, or make the connection between how a new technology used for one project might address a need in another, enables me to strategically drive our work forward." This is true, she adds, of understanding the needs and skills of the team members and partners she works with. "That, paired with an interest and some ability in communicating technical and scientific concepts, has equipped me well for the position I am in today."

Although she admits that leading people and projects is a better fit with her personality than coding or statistics, Regan strongly encourages young women to equip themselves with technical skills. "One of the best pieces of advice I received when pursuing my graduate degree was to focus on building hard skills during my education and early career—taking the difficult science and technology classes, learning GIS and basic programming, taking advanced statistics instead of the 'light' classes that sounded interesting and fun," she says. No matter what your passion or your future career path may be, she says, **"if you want your efforts to have maximal impact on the world—be that building a more sustainable planet, promoting more equitable social systems, or even just making beautiful art—focusing on STEM fields early in your career will empower you with the tools to do so."**

For Regan, the beauty of GIS technology is that it allows information to be visualized in a compelling way that makes another person see or understand something that was inaccessible to them before. **"I love the moment when you put a new, beautiful map up on a screen and the room goes silent as people eagerly lean forward, squint their eyes, and drink in what is newly revealed,"** she says.

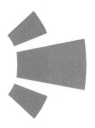

On a day-to-day level, Regan says, she is motivated by the people she works with: "Being part of a mission-driven hardworking team, and even more so, having the responsibility of leading such a team, is profoundly motivating." On a deeper level, though, what motivates her today is largely the same as what motivated her younger self after sailing around the globe: feeling awe at the diversity of life on Earth and a commitment to its wise stewardship. Her work, she says, is "inspired by a sense of wonder in the natural world and desire to keep it functioning for the benefit of my children, their children, and everyone else who comes after us."

PATRICIA SOLIS

Serving as an ambassador for people, places, and peace

Position

Research associate professor of geography
School of Geographical Sciences and Urban Planning at Arizona State University (ASU)

Executive director
Knowledge Exchange for Resilience at ASU

Education

PhD in geography
University of Iowa

MA in geography
Kansas State University

BS in physics and BA in German literature
Kansas State University

ONE OF THE BIGGEST THRILLS of Dr. Patricia Solis's life was receiving a diplomatic passport to travel in Latin America. She had been appointed to the Organization of American States (OAS) Pan American Institute of Geography and History (PAIGH), an organization promoting scientific collaboration throughout the Americas, where she served first as vice president and later as president of the Geography Commission. When she got the passport and saw the OAS embossing on it, she grinned from ear to ear. "I kept thinking how my younger self would think that I, as my older self, was so cool to speak two foreign languages (Spanish and German), travel the world, and actually hold a diplomatic passport," she says. "When I was younger, I dreamed about engaging with people in all of these amazing places on planet Earth, and I get to do that now."

Small-town opportunities

Growing up in the small town of Neodesha (population 3,400) in southeast Kansas, Patricia and her twin sister, Jennifer, were both drawn to science and math. "I always knew I wanted to be a scientist of some kind," Patricia says. "I loved math and working with computers. I also really loved teaching about computers and loved learning about places and people from all over the world, so I guess it's not

Patricia at PAIGH headquarters in Mexico City in the gallery of member flags.

a surprise to anybody who knew me then that I'm now a professor at a university and a researcher."

In school, her teachers encouraged her to take as many math and science courses as possible. In fact, they ran out of courses, and in high school, Patricia had to drive to the next town to take classes from the community college so she could continue doing what she loved.

Their parents always told the twins that they could do anything and be anything they wanted. **"In our family, we are [all] girls, and my dad never discouraged us from doing anything or learning anything,"** Patricia says. "He taught us everything he knew about fixing things in the house, like plumbing and cars and life in general. My mom was a legal secretary, always super organized, and imparted to us amazing management skills and attention to detail. I think these experiences have helped me manage my career, advocate for myself, and take life into the direction that I wanted without feeling intimidated by anything."

Patricia and her twin sister, Jennifer, through the years. (Patricia is on the right in the first photo and on the left in the other two.)

Public university and study abroad

When it was time to apply for college, Patricia received many offers of admission and scholarships, but the costs made it impossible for kids of working families to go anywhere but a public state institution. Yet she is glad she did because she loved her time at Kansas State University. "Institutions like public state universities are very special to me because of their mission to support students who may not be able to afford private, expensive schools," she says. "The mission of land grant universities to serve the public resonates with my own vision for my life. At the same time, these places also gave so much space for me to grow."

At Kansas State, Patricia received scholarships to study mechanical engineering. She loved the coursework and took many electives and language courses to satisfy her love of connecting to people around the world. Two years into college, she had to change her major to physics because the degree plan for engineering did not allow her to continue also taking courses in music and languages, which she wanted to do.

As a physics major, Patricia was the only female student in her classes, yet the small class size and involvement of the professors made Kansas State a supportive environment. Still, most physics majors went on to focus on nuclear physics, but she didn't know what to do with that degree for a career, so she escaped for a year on a study abroad scholarship program to Switzerland.

There, she studied at ETH Zurich (Swiss Federal Institute of Technology in Zurich), a public research university that Albert Einstein attended, and where Patricia says she was inspired and humbled as she worked to perfect her German. She stumbled into a geophysics course, and then took environmental geography and remote sensing courses for the first time. "I remember walking down the hallway of the Geophysics Department, and the walls were covered with incredible satellite imagery that made it look and feel like an art gallery," she says. "I was fascinated and spent so much time trying to understand what we could learn from those images. **Incredibly, in 1992 in Switzerland, they were using satellite Earth observations, GIS, and fractal geometry to identify and predict avalanches in the Alps. Once I understood that this technology—this beautiful art and science—could actually save lives, I was hooked.**"

When Patricia returned to the US, she immediately finished her physics degree, added a German literature degree, and then started a master's program in geography, taking on a graduate assistantship with Kansas State's International Community Service program to help other students gain the same kinds of experiences. **"I was committed then and there to working around the world in collaboration and solidarity with other scientists to use these wonderful tools for mapping for resilience, humanitarian, and development purposes,"** she says. It is a commitment that Patricia has stuck to ever since.

Beginning her career

After finishing her master's at Kansas State, she took a job at a small city in western Kansas, working in economic development and eventually starting a new community development office that took on local issues such as education, jobs, beautification, and historical preservation. It was a great experience, she says, although difficult at times, working with local government and workplace politics. Yet she gained valuable insight into how cities function and how local communities work together across the public and private sectors.

After a few years, she accepted a fellowship from the president of the university to get her doctorate in geography at the University of Iowa. It was there, while working at the Iowa Social Science Institute, that she met and married her husband, Dario, of Panama. "I was thankful for the fellowship because, in the final writing year, I was very sick with my first pregnancy, and it allowed me to spend the time on finishing my research. As my belly grew, so did the word count on my dissertation," she says. "I submitted the first full draft and went into the hospital to deliver the following week. Through this experience, I knew I wanted to do research and continue working with geospatial technologies and environment and development issues, but I was not so sure that I wanted to be a professor at that time. It seemed quite lonely writing your dissertation largely by yourself, and I couldn't imagine not working more closely with people and projects."

A job that fits

At that point, the Association of American Geographers (AAG) was going through a major leadership transition, and Patricia happened to see a job ad in the newsletter. "I read the ad with my husband, and we decided that even though I had just had our first son and was about to defend my dissertation, it was just too good a fit not to try," she says. "So, our little family, with the three-month-old in tow and a freshly approved PhD, was off to Washington, DC, to begin what became an exciting 12 years. Learning, at a national level, about everything that geographers, GIS, and related scholars and practitioners were doing enriched my thinking, grew my

> # Fun fact!
>
> **What boosts her well-being:**
> "Swimming. I love the feel of the water and the calm that it offers. Well, it's also very fun to play with my kids in the pool."

Patricia at an AAG annual meeting posing with the group's renowned blow-up globe. She has attended every year since 2002 without interruption, although it was held virtually in 2020 and again in 2021 because of the pandemic.

Patricia and her husband, Dario, at the Panama Canal.

network, and gave me a purpose within a community that I care deeply about."

While in that position, Patricia and her family moved to Panama, for family reasons, right after her second son was born. Since AAG united a broad national membership, she was able to work remotely and, in fact, used that experience to launch programs to engage geographers in international and developing regions. She lived in Panama for 10 years and began telecommuting in 2005, so the recent challenge of working from home and meeting remotely with colleagues is familiar to her—only this time, during the COVID-19 pandemic, everyone has been struggling through it together.

"I am particularly proud of the work that I did at AAG to advance international participation in the organization and the programs for youth around the world, like My Community, Our Earth, for which I had to also write competitive grants and win and execute external funding to keep going," Patricia says. "I also initiated a series of National Science Foundation–funded projects with the goal to broaden participation and diversity in geography departments in the US and build upon the intellectual value of including geography as a scientific discipline." In this way, Patricia became proficient at building funding streams for research and engagement projects—a skill that has been useful to her no matter where she's worked.

Mapping with the next generation

Despite her other commitments, Patricia never fully gave up on teaching pursuits and, to keep herself at the top of her game, secured an adjunct affiliation with George Washington University—teaching at night when she lived in DC and, in Panama, coteaching study abroad courses with respected colleagues.

In 2014, Patricia transitioned fully to academia, thanks to an opportunity from one of her advisers from Kansas State, who was then president of Texas Tech University. Texas Tech gave her a platform to start YouthMappers, with continued funding from the USAID GeoCenter. "YouthMappers, like all of my favorite activities, is built upon close collaborations with colleagues" from other institutions, Patricia says. She still directs YouthMappers, a consortium of student-led and faculty-mentored chapters at more than 240 university campuses in 55 countries as of 2021, which works together with USAID to create and use open spatial data for humanitarian development purposes. YouthMappers sponsors dozens of projects around the world, in which mapping supports decision-making and understanding of everything from health-care access to flooding to disaster response. YouthMappers also offers annual programming in fellowship cohorts, where students from many countries come together to learn from the team and from each other and grow their solidarity around open, humanitarian geospatial action.

Patricia gazes at a giraffe during a safari in South Africa, as part of a field trip for the 2019 YouthMappers Leadership Fellows workshop.

YouthMappers grew out of Patricia's foundational ethic of investing in the next generation. **"Success, to me, looks like the eventual success of other people who have been a part of my work,"** she says. **"It means so much to see those seeds that were planted grow and flower.** For example, one high school student participant in a US State Department GIS international program later went to university, and then launched a youth environmental movement. She was kind enough to reach out to me to let me know that she took some of her inspiration from the experiences we had. It's a privilege to see those things happen."

She says the women she has engaged with go on to inspire her and many others. One young woman from one of her fellowship programs, who grew up in the Maasai community of Kenya, went to work for NASA in the US and developed geospatial algorithms to help her tea-growing community predict frost and protect crops. Another young woman from Brazil had won a map poster competition, and YouthMappers supported her attending the Rio+20 side

From left, Patricia and two of her foremost YouthMappers collaborators, Astrid Ng and Marcela Zeballos.

event in 2012. The young woman ended up graduating with her PhD and recently rejoined YouthMappers and is now a regional ambassador, working across Brazil to pay forward those opportunities. Patricia says she is awed at how these women have this kind of impact and that YouthMappers keeps growing and affecting the lives of young people.

Working to improve the world

In 2018, a geographer colleague reached out to Patricia and encouraged her to apply for a new project funded by a local philanthropy at Arizona State University. Again, when she read the job ad, she realized that it was one of those rare opportunities when the position is a perfect fit. Her current position is research associate professor of geography in the School of Geographical Sciences and Urban Planning at ASU, where she serves as executive director of the Knowledge Exchange for Resilience. Its mission is to meet the resilience needs of the local community with the campus-wide intellectual resources of ASU in data, skills, partnerships, and insights.

For Patricia, as a broadly trained geographer, educator, and applied researcher, with specific data and technology expertise, it was the opportunity of a lifetime to meld collaborative scholarship with her experience in teaching and leading many types of audiences. "I am sometimes stretched by the range of activities we are tasked with and the vast network of people who are engaged in this endeavor," she admits, "but it is exhilarating, and I cannot think of a better place for me to call my career home. **Sometimes it can get overwhelming, and that's my biggest challenge, but I'm continuously learning, connecting, and working to help real people in lots of places to address problems with science and evidence-based responses.**"

Patricia works on projects involving food security, economic insecurity, health and well-being, infrastructure, and housing and shelter. In Arizona, she helps make hidden vulnerabilities visible by mapping where the needs for utility assistance are greatest—for instance, among people in mobile homes who suffer excessively in the heat. **"Many people think of using GIS to get answers, but I also like to use it to figure out what are the right questions to be asking,"** she says.

Internationally, she has worked on a project to speed up the identification of features for rural electrification projects, focusing on the need for women and girls to have access to electricity for household labor—a need that may be overlooked if electrification is concentrated on businesses instead. She works with many people on this project, including local students in Sierra Leone as well as technical staff from Facebook and Mapillary and electrical power engineers from ASU to use artificial intelligence to quickly identify where people live who need access to electricity. "My favorite part of the job is the people," Patricia says. "I absolutely love working with people and technology but people first. Unfortunately, I seem to always run out of time to do everything that I can think up doing. **This kind of work, especially in humanitarian circles, is so much in demand. There's just too few of us geographers to address all the world's problems and all the opportunities to make a difference in people's lives."**

The best advice that Patricia has received is to pay it forward. **"When you invest in other people, you build yourself up, too. It's the best way to show gratitude for what others have done for you, even if you don't know who they are.** For example, I have no idea who set up the scholarship fund for me to study abroad those many years ago. I just keep paying that forward. Especially the attention, time, and energy that you put into collaborating with peers and younger people is like an investment. I know I've received many dividends from this investment over the years."

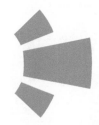

Fun fact!

"Because of my husband, Dario, who is from Panama, I became fluent in Spanish and learned how to cook Panamanian dishes—people tell me how authentic the food I prepare tastes, and I love hearing that. Dario and I enjoy dancing salsa and bachata together."

Patricia in La Paz with a traffic zebra. Traffic zebras—people in zebra costumes—help people cross the street.

She encourages young women to develop a solid professional network, such as a board of advisers, because no one single mentor is going to be able to give you everything you need. Reach out to your peers as well, she suggests, and support each other.

Patricia keeps a certificate in her office that is especially meaningful to her. In 2017, she attended a professional conference in La Paz, Bolivia, called Geographies of Peace (playing with the meaning of La Paz). Along with the pleasure of being in a beautiful city and reuniting with so many of her colleagues—it was so much fun, she says—she was honored to be one of the key speakers awarded honorary citizenship to the city. She appreciates the peace-loving designation and sees it as a validation that she is a bona fide citizen of planet Earth who enjoys being a scholarly ambassador for knowledge. ◢

NAVYA TRIPATHI

Pioneering the future of GIS

Position

Student

Education

Plans to pursue a degree in the medical field

NAVYA TRIPATHI is only in the eleventh grade, but she's already using GIS like a pro. For the last three years, the teenager has been working on a project studying the demographic and spatiotemporal analysis of drug overdose deaths in the United States on a county scale. "The drug epidemic is widespread in our country and claims a lot of lives, which can be prevented by early intervention. This topic is impactful, and knowing that excites me to keep furthering my project to learn more," Navya says. She's currently working on incorporating the spread of COVID-19 into the drug overdose study to consider the pandemic's impact on drug overdose deaths on a geographic scale.

According to Navya, "GIS found me more than I found GIS." The daughter of two parents working in GIS, she was introduced to it at a young age in the form of brightly colored maps. "As I got older, the maps began to appeal to me for reasons apart from their aesthetics. I was intrigued at how easily maps could convey a message on a topic, to experts and laymen alike," she explains. "I believe the communication of ideas is necessary for the society to move forward, and something that can speak to everybody, like maps, is an extremely powerful form of communication. That's what grasped my interest, and through all my projects, I hope to improve my communication skills by creating visuals for imminent problems in society."

Navya at the beach, one of her favorite places.

Fun facts!

"I love singing and dancing and generally always have music playing around me."

Fun thing on her desk: "I have a photo album with pictures of my friends and family on my bookshelf and a bullet journal (with more stationery than I could ever use!) to keep me organized."

Favorite thing about GIS: "Ability to amass enormous amounts of information to tell a story through a single graphic."

She's also been fortunate, she says, to attend conferences with her parents and watch them present—something that's developed her own interest in science and public speaking. That comes in handy, Navya says, now that she's started presenting at conferences herself. In 2019, she was selected to present at the Urban and Regional Information Systems Association's URISA GIS-Pro conference in New Orleans, Louisiana, as the youngest presenter in the moderated presentation session for college professors. "I truly enjoyed the experience and learned a lot from listening to other presenters," she says. She'll also be presenting at the 2021 Law Enforcement and Public Health conference and hopes to be selected as a presenter for a future Esri User Conference.

Embracing GIS

Although she's gained a lot of experience since then, Navya presented her first GIS project at the URISA high school poster competition in 2017, where she mapped Hurricane Irma's impact. The project was well-received, and she won first prize, which included an ArcGIS software CD. "That not only gave a big boost to my interest but also gave me a vehicle to pursue it," she says.

Robert Thomas of the American Red Cross, where Navya has volunteered since 2018, has also helped Navya, giving her the opportunity to gain more GIS experience and participate in its real-world applications while serving the community. And her mentor, Dr. Nancy Hardt, professor emeritus at the University of Florida, has guided her through many of her projects. "The idea of becoming successful and being able to contribute to the society motivates me to work hard every day. I want to give back to those who made me who I am, and that motivates me to do well," Navya says.

Navya's favorite part of working with GIS is playing around with different data layers and seeing how the map changes to show different relations by correlation. "I find it intriguing to see how factors that are either correlated or confounded can be depicted on a map, and how certain depictions of the data can change the meaning of the map," she says. **"I also like how easily a map can combine several features, and GIS makes it easy to analyze and understand for everyone."** She's not as big a fan of data collection and cleaning, though she acknowledges the necessity. "It is a cumbersome process," she says, "but nevertheless the most important one as data availability and data quality are ultimately what guide a study."

Navya in first grade.

Navya at the URISA GIS-Pro 2019 conference, where she was the youngest presenter there.

Navya in Las Vegas for the Mu Alpha Theta National Convention.

Pursuing knowledge

In addition to her GIS work, Navya is part of the prestigious and nationally recognized Buchholz High School Math Team, which has been Math National Champion for 12 of the last 14 years. She was also one of the recipients of the Duke Talent Identification Program national award, which recognizes high-achieving SAT scores in the top 1 percent of seventh graders. "I have been fortunate to have a strong family that supports me in everything I want to pursue. In addition, my school and friends' support all along has always provided me the moral support and the confidence I needed to keep me focused," she says.

Navya also credits her schooling with encouraging her academic interests and passions, saying, "**I love learning new concepts and seeing how these concepts give me new perspectives on my own projects. It makes me realize how much knowledge there is in the world, and how everything affects everything else.**" Education, especially accessible education, is a cause Navya is passionate about. "**Knowledge is one of the most powerful assets in the modern world, and this applies to intellectual and creative knowledge,**" she says. "I also believe that altruism is important and see it as a human obligation to help others. I feel that is one thing our world could benefit greatly from."

Looking to the future

The application of GIS in the medical field—specifically, epidemiology—is a big interest of Navya's. She's always had an interest in the medical field (and GIS), but her project on the drug epidemic combined them. Her goal is to go into the medical field, although she hasn't decided on a specialty and intends to gain more knowledge in the coming years. "I have immense respect for medical professionals, as their job is a selfless one," she says. "They work long hours, and with each patient they treat, they impact a whole

Favorite trip: "My favorite trip was when I went to Las Vegas for a week with my friends (Buchholz Math Team). We went to compete in the Mu Alpha Theta National Convention and won the national title. I learned a lot on that trip and brought back priceless memories."

On her bucket list: "I love the beach, and while I am surrounded by beaches in Florida, I have always wanted to go to the Maldives to visit the beaches there."

What boosts her well-being: "I enjoy listening to music and spending time with my friends and family. When I want to do something on my own, I like to cook, make jewelry, or go for a run."

Navya at the Alachua Region Science and Engineering Fair in Gainesville, Florida, in 2019.

family. I am fully aware of the commitment and dedication needed to pursue a career in the field of medicine, and I believe I am up for that challenge. It is my belief that it is a profession where the earning is more in good wishes than in dollars, and that's what matters in the end," she explains.

Through hard work and perseverance, Navya has accomplished much in her young life. "My greatest strength is my family support and teachings, without which I would not be what I am today," she says. She has high hopes for the future, and although she says she can be a bit of a perfectionist, she's learning to respect and appreciate the imperfections in life. To her, success is "giving my best to achieve the goal that I have set and **to be content with the outcome, learning from the process and using the learnings to improve in the future with the hopes to better the outcome."**

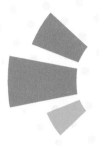

KALPANA VISWANATH

Pinning her business on the safety of cities

HENEVER DR. KALPANA VISWANATH visits a new city, she finds time to sit in a park or at a roadside café to people-watch and observe how people engage with the city and its spaces. She has lived in megacities all her life—including Bangkok, Thailand, and Chennai, Mumbai, and Delhi, India—and, she says, "Despite all the challenges, I love them for their vitality and energy."

She is passionate about building more equitable and caring cities. She is also passionate about gender justice and equality. To translate these passions into "change on the ground," she has cofounded an organization that maps cities and public spaces for safety, accessibility, and inclusiveness.

Kalpana cofounded Safetipin with her husband, Ashish Basu, a technology design expert. The idea for the app came up over many discussions, and even dinner table arguments, that the two cofounders have had over the years. The aim was to use technology, data, mapping, and urban design to "build a world where everyone can move around without fear, especially women," Kalpana says. The organization offers a free crowdsourced map-based app called My Safetipin that helps users make safer decisions about their mobility, based on the safety score of an area. At the core of the app is the Safety Audit, which is a tool to analyze a given area on the basis of physical and social parameters, such as lighting, openness, visibility, people, security, walk path, public

Position

Cofounder and CEO
Safetipin

Education

PhD in sociology
Delhi School of Economics, India

MA in sociology
University of Mumbai, India

transport, gender usage, and feeling of safety. The app also suggests the safest route, which may not necessarily be the shortest. Safetipin Nite is an app that collects photographs of cities at night, which are then analyzed manually and with machine learning, to supplement the crowdsourced data on My Safetipin.

Kalpana and Ashish started Safetipin in 2013, and since then, she says, she has "experienced such a great (and very challenging sometimes) learning curve." Kalpana's background is in sociology, but she now heads an organization that works with technology, data, mapping, and urban design—all of which she has learned on the job. She finds that she enjoys leading a team of people with diverse backgrounds—technology developers, urban planners and designers, engineers, and social scientists. And she loves traveling around the world, building partnerships, and meeting people doing meaningful and innovative work to address the challenges of urbanization and gender equality.

Kalpana speaking at an event in Delhi, India.

Going from place A to B but the fastest route isn't the safest?

Use our Safest Route feature to take the route with the highest safety score.

SAFETIPIN
Supporting Safer Cities

The My Safetipin app.

Kalpana at the New Cities Summit in Delhi, where she was awarded the 2017 Global Urban Innovator award.

Though Safetipin is still a young organization, it has worked with more than 15 city governments in India and other developing countries; is currently developing mapping projects in Colombo, Sri Lanka, and Durban, South Africa; and has won several awards for leadership and transformational technology. Yet, for Kalpana, the biggest achievement has been "to create this organization that is continuously learning and innovating."

A multicultural education

Growing up in Bangkok, Kalpana didn't have a clear idea of what she wanted to be. She went through a phase of wanting to be a waitress, then a teacher, and even a flight attendant. But, she says, "The one thing I knew from around age 13 was that I wanted to do something that had a social dimension." In Bangkok, she attended an international school, which gave her wide exposure to people from different backgrounds. "Being in a multicultural space gave me the opportunity to see, appreciate, and embrace diversity," she says. "I believe this school experience has made me the liberal and reflective person that I am today." She was also a voracious reader and drawn to books that examined women's lives, which made her a feminist from a young age. Also, she says, "My parents treated me equally and allowed me to dream."

Kalpana went on to obtain a master's degree and then a PhD in sociology, which allowed her to hone her skills in research and

Tip!

"The best piece of advice I got early in my career was to keep looking down and sideways even as you move upwards in your career. This has helped me understand different perspectives and keep helping other women up as I was helped and continue to be helped."

critical thinking. As a student, she was involved in student movements and the women's movement, which, she says, "developed my political and social self." She realized that, although she loved doing research, she wanted to work in an arena where she could be a practitioner and bring about concrete social change. With this goal in mind, she began her career working for an NGO, Jagori, that focused on women's rights. Her role there was to develop a program on building safe cities, which, she says "forced me to work with a wide range of people and especially taught me that working with governments requires patience and a long-term vision."

After working with the City of Delhi to map urban spaces for safety and accessibility, she consulted with UN agencies to take the model of safety mapping to cities around the world, including in Asia, Africa, and Latin America. **"It has been wonderful to work in and learn about so many cities,"** Kalpana says.

Converting ideas into communication

Kalpana believes that her greatest strength is her ability to communicate and build a strong network. In many cases, presenting at a conference has led to a partnership to work in a new city or context. For instance, in 2013 she went to Bogota, Colombia, and spoke at an event, which led to the city government adopting My Safetipin, translating the app, and working with the organization. A year later, in Jakarta, Indonesia, she had a brief interaction with the governor; he liked the app, promoted it on the media, and My Safetipin got more than 4,000 downloads over a single weekend. **"I love converting ideas and thoughts into communication,"** Kalpana says. She used to write a weekly column on urban spaces for a major Indian newspaper, which allowed her to communicate her ideas to a wider audience. And, with My Safetipin, she says, **"Maps can be used to tell stories. We have used the [ArcGIS StoryMaps] tool to bring the spatial data alive by including people's voices and experiences of those spaces."**

Kalpana during a women's festival at a newly pedestrianized road in Delhi, India.

Reflecting on work-life balance, Kalpana acknowledges that women often must face the challenge of building a career and a family at the same time. She believes it's important "not to try to be a superwoman and put too much pressure on yourself," she says. In her early career, she tried to work with flexibility so that she could also look after her children. But she was 48 when she started her company, and her children were grown by then, freeing her to work "without too much guilt and conflict," she says.

Today, though, she believes that young women **"are living in a time where they can dream and put their needs first," she says. "This is, in some sense, unprecedented in history, as women have always been taught to subsume their own needs."** Her hope for young women, whether they are considering entering STEM fields or something else, is that they will "think out of the box and not fear doing something different." In launching the My Safetipin app and adopting innovative technology to create a safer world for women, Kalpana says she is driven by "the belief that we are social beings and must learn to live together with compassion. **My motto is that people working with commitment and compassion can change the world in small and big ways."**

JULIA WAGEMANN

Expanding the network of female leaders in GIS

𝒥ULIA WAGEMANN GREW UP in rural Germany, on a farm in a small village with fewer than 50 inhabitants. It was not, she says, the type of environment that encouraged a young girl to dream big. As a teenager, she'd already rebelled against the life that was expected of her; she dreamed of moving to a larger city, having an unconventional career (perhaps as a travel journalist), and seeing the world. But, she says, "Since no one around me could guide me with that, I needed to try out different things and make my own experiences."

Between graduating from high school and starting university, Julia spent three years "trying out things." After internships in the media business, she traveled to Australia, where she worked for a year as a diving teacher on a diving boat at the Great Barrier Reef. Then she worked for a season as a travel guide in France. These travel experiences, where she learned more about the unique Great Barrier Reef and mountain ecosystems in general, also fostered her interest in geography—an interest that led, ultimately, to a career in geospatial data and systems.

Finding her passion

For the past six years, Julia has worked at the intersection of data providers and data users to make large volumes of open meteorological and climate data more accessible and used. She works as an independent consultant, developing

Julia working on climate data visualization at the European Centre for Medium-Range Weather Forecasts.

educational tutorials and workflows with the European Organization for Meteorological Satellites (EUMETSAT). For the European Centre for Medium-Range Weather Forecasts (ECMWF), she coordinates the ECMWF Summer of Weather Code (ESoWC), an online program she implemented to develop innovative open-source climate-related software. In her free time, she works on a project to make Copernicus open climate data available to Google Earth Engine users. She's also a PhD candidate in geography and computer science at the University of Marburg, Germany. What drives her, Julia says, is her dedication to making "large volumes of Earth observation data better accessible for everyone" and seeing the gains in knowledge that result.

Julia's fascination with geospatial data, mapping, and technology began as an undergraduate at the University of Marburg, where she learned data analysis skills in introductory courses on GIS and remote sensing. As the data volumes grew, she had to learn programming as well. She takes particular pride that her undergraduate thesis was recognized as the best by the Faculty of Geography at the University of Marburg in 2011: "As a first-generation university student, I navigated through university without advice, and this award showed me that I did it, and I finally found my passion."

As an undergraduate, she spent two semesters studying abroad, at Université Laval in Québec, where she took all the courses and exams in French and signed up for a coding class without having coded before. "This was certainly a year where I put myself out of my comfort zone. The hard work was rewarded, though, as I finished all courses with straight A's, and at the same time, I started my journey on learning how to code," she says. That journey continued through her master's program in geoinformatics at the University of Marburg. Developing her coding skills was an ongoing challenge—for a whole year, she spent every weekend working on coding assignments. But, she says, **"Looking back, this time investment still pays off today, as knowing how to code is the biggest asset of my skill set."**

Having forged her own path to success, Julia has these words of encouragement for young women considering a career in STEM or GIS: **"You can do it! Do not let anyone tell you what you can or cannot do, and school is not the only place to find your interests. There are many ways that lead you to learning about GIS, mapping, geospatial data, and coding. Embrace change and push yourself from time to time out of your comfort zone, as these times are the ones where you grow and learn the most."**

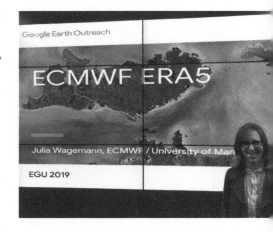

Julia presents work bringing a subset of ECMWF Reanalysis v5 (ERA5) climate reanalysis data into Google Earth Engine.

Confronting bias

Along the way, though, Julia has come up against some hard truths about lingering gender bias in the field. The first time she encountered it was in high school, when she changed schools so she could do her A levels in math and physics. But, at the new school, out of 28 students, she says, "we were only four women, and the math teacher, who was male, often took the opportunity to make us women look like fools."

The next time was at the European Space Agency (ESA) in Italy. Right after she defended her master's thesis, she was offered a Young Graduate Traineeship position at ESA—a big opportunity she'd never thought she would have. Her time there taught her how the European space industry works and helped her develop her professional network, but as a job, she says, "Unfortunately, it was a disappointment." She wasn't involved in any real projects, and her supervisor turned out to be a poor role model as a mentor and manager. It was also, for Julia, "the first time realizing that there is a glass ceiling for women. In upper management, there were hardly any women at ESA."

By contrast, after 15 months at ESA, she got a job offer from ECMWF, where her supervisor was supportive and gave her enough trust and freedom for her to grow professionally and develop her skill set. Spurred by these experiences, and recognizing the importance of mentorship, in 2019 Julia cofounded the Women in Geospatial+ network, a professional network to promote gender equality in academia and the geospatial industry. Since its inception, the network has grown to over 2,400 members worldwide and offers many activities, including a speakers' database, a webinar series on professional career development, and a mentorship program.

Julia moderating a panel discussion as part of the Women in Geospatial+ network.

Tough choices

In her personal life, too, Julia has faced some tough choices. To take the position at ECMWF, she had to move from Italy to the United Kingdom, and for two and a half years, she and her husband had a long-distance relationship between the UK and Italy. Then, in the spring of 2017, she got pregnant; at the same time, she and her husband both received very good job offers. Julia had an offer in Germany, and her husband had an offer in Italy. "As I was pregnant, it was clear for us that we now had to decide to live in one place," Julia says. "It was a decision on whose career we should put first, and it was very tough," she says. In the end, they decided to move together to Italy. But, says Julia, "the decision catapulted me into a position I never wanted to be in. For the first time, I moved to a country without a job, became a mum for the first time, and left a job and workplace (at ECMWF) with which I strongly identified myself."

At the time, the situation might have seemed daunting. Three years later, though, Julia is confident that it was "the best decision we could have made." As she explains, "This situation put me quite heavily out of my comfort zone, so I needed to look for new opportunities." She started working on a PhD, a step that had long been recommended to her. Her doctoral research involves user requirements of cloud-based data services. At the same time, she has had the opportunity to work as an independent consultant. "This allows me to get to know other organizations and requires me to learn a new set of skills," Julia says, quoting one of her favorite mottoes: "Growth and comfort do not coexist." In fact, Julia believes that her greatest strength is "putting myself constantly out of my comfort zone and a desire to keep my learning curve steep. Also, the ability to think that there is not a skill I cannot learn. If I need this skill to get where I want to go, I will learn it."

Julia climbing Iceland's highest peak with her husband in 2016.

Julia draws strength and courage from her sister, who was paralyzed in a skiing accident at age 23. "Her resilience and optimism to make the best of her fate up until today is just inspirational."

In addition to her quest to make weather- and climate-related data more accessible, Julia is passionately committed to the cause of "equality and diversity in the geospatial industry and academia." Since she never places limits on herself, she says, **"it is heartbreaking to see that there are many women starting out in the geospatial field, but only a few will be able to be in a leadership and decision-making position. There is certainly a gap we have to solve,** and through Women in Geospatial+, we want to change the current status quo by creating a strong network of women+ leaders and change makers." It's an ongoing challenge but perhaps not quite as steep as another challenge Julia has set for herself: summiting a "5,000er"—a 5,000-meter mountain peak.

FAUSTINE WILLIAMS

Improving health outcomes for underserved populations

Position

Stadtman tenure-track investigator
National Institute on Minority Health and Health Disparities

Education

PhD in applied social science
University of Missouri, Columbia

MPH in public health
University of Missouri, Columbia

MS in health informatics
University of Missouri, Columbia

BA in information science and psychology
University of Ghana, Legon

BS in environmental health
School of Hygiene, Accra, Ghana

"WE CANNOT SOLVE the issues of minority health and health disparities without increasing the number of minorities pursuing these careers," says Dr. Faustine Williams, a Stadtman tenure-track investigator within the Division of Intramural Research of the National Institute on Minority Health and Health Disparities (NIMHD) who is working to improve the health and well-being of racial and ethnic minority and underserved populations. As an immigrant Black woman, it's a cause close to Faustine's heart and one that motivates her every day.

"Success to me is using my knowledge, skill sets, and position to make sure the voices of racial and ethnic minority and vulnerable populations are not silent but are heard, as well as helping improve health outcomes and reduce health disparities. I always joke with my lab that our goal is to help reduce or eliminate health disparities in the society and win the Nobel Prize," Faustine says. Her lab, the Health Disparities and Geospatial Transdisciplinary (HD and Geo-TransD) Research Program, nicknamed the Fancy Methods Lab, focuses on health disparities and geospatial transdisciplinary research to understand factors influencing health disparities and develop effective interventions to improve health outcomes among minorities and underserved populations. She uses mixed methods, including community-based dynamics and participatory research and GIS to understand the complex interactions of factors contributing to the health of minority populations and health disparities.

Among the numerous projects her lab is currently working on are (1) mining Twitter data to learn about the impact of COVID-19 on minorities and disadvantaged populations; (2) the impact of COVID-19–related social-distancing policy and measures on mental health as well as other long-term health problems with the ongoing disruptions in health care and other services; (3) the association between neighborhood cohesion and sleep outcomes in immigrants; (4) the differences in self-reported hypertension and cardiovascular diseases among African American, African-born, Afro-Caribbean–born, and Latino immigrants in the United States; and (5) spatial access to grocery facilities and cardiometabolic condition.

Faustine says she is motivated by her work because she wants to see people "healthy and happy."

As a 2018 National Institutes of Health (NIH) Distinguished Scholar, Faustine also participates in a cohort model for enhancing diversity and inclusion of principal investigators in the NIH intramural research program, which provides mentoring and professional development activities to foster research and promote scientific workforce diversity. "My goal is to build a well-established transdisciplinary research program with other NIH intramural scientists," Faustine says. **"My ultimate long-term desire is to build a long and productive career, working toward helping to eliminate disparities so that everyone can benefit from good health,"** she explains. "But there is still a lot of work to do. I am grateful to NIH/NIMHD for affording me the career opportunity to carry out my dream of helping people."

Finding her path

Born and raised in Ghana, West Africa, Faustine has been working toward a career helping the underserved since she was 13 and experienced her first taste of public health awareness. In the

Tip!

"Motivation and the passion to not quit when the 'going becomes difficult,' as we say in Ghana, is important. Always remember, winners never quit and quitters never win."

mid-1980s, *dracunculiasis*, or guinea worm disease, a parasitic infection that comes from drinking unfiltered water contaminated with guinea worm larvae, became rampant in Ghana. Faustine was the youngest to participate in a task force assembled by the World Health Organization and the International Task Force for Disease Eradication. She traveled to villages, providing health education on water filtration and other preventive measures. **"From bringing extra food to school for those less fortunate to handing out clothes to those in need, I did not know the good Lord was setting me in motion for what would quickly become a shift toward a future career in research to help the underserved,"** she says.

As a child, Faustine was captivated by science and anything that involved calculations. She was taught by her father, a mathematician, because she was not able to attend school until she was six. She even dreamed of becoming a mathematician like him. Once she started school, she says, she never wanted to leave and couldn't wait to return after the holidays. She attributes some of that to her father's not allowing her to explore anything outside her studies, which she somewhat laments since she was a good athlete and was never able to pursue sports as fully as she would have liked.

After high school, Faustine worked for several years before eventually attending school for environmental health. She went back to work for a few years but was still not satisfied and pursued further studies. In 2005, she was admitted to the University of Missouri, Columbia, graduate school, where she would eventually receive two master's degrees, a doctorate, and a graduate certificate in GIS and community development.

But she almost didn't attend graduate school. A travel agent in Ghana took her money and disappeared without giving her the plane ticket. Although she was 1 of 10 women in the world awarded the American Association of University Women (AAUW) International Fellowship Scholarship, she worried that taking another $1,500 from her savings for another ticket would

Faustine, left, with her brothers Paul and Prosper and her cousin YaaYaa, second from right, at church in Ghana in 1989.

Faustine at her graduation with first-class honors from the University of Ghana, Legon, in 2003.

make it impossible to survive in a foreign country. Her mother, who never finished elementary school after having to drop out to care for her brother, found out what happened and took Faustine to a family friend for a loan. He refused. "My mom went on her knees and pleaded with him until he changed his mind (she even rolled on the floor). I might not have been here today without her selfless love to do whatever it took to see her daughter educated and achieve my dreams," she explains.

Rising above

Faustine is no stranger to hardship. She believes that every circumstance, good or bad, presents an opportunity to rediscover herself. Hence, her life motto: **"All things will work together for my good, even the bad ones." Faustine adds, "I always say, no matter how many times I fall, I will rise up and restart if that is what it takes."** It's a truth she's found evident throughout her life—nothing can stop her except herself.

As the only African or Black person in her master's course, Faustine faced jokes and suggestions that she wasn't there on merit but because she was African. Because of her accent, it was even worse. "People had already made up their mind that they could not understand even before I opened my mouth. Even to this day, some people treat me this way," she says. Naturally quiet and introverted, she quickly learned that she had to do something if she wanted to succeed. Nobody wanted her on their team for group work until they realized she was smarter than they were. **"All of a sudden, everyone wanted to work with me," she says. "This started to boost my confidence. I am still the same Faustine ... but it does not worry me if some people judge me based on my appearance or accent. I am proud of my background/heritage and who I am."**

Even now, as accomplished and successful as she is, Faustine still faces people assuming she's not intelligent because of how she looks and speaks. "Sometimes after a presentation, people will give me a compliment and say things like, 'Wow, you are smart. You know your stuff. Did you know English when you came to America?' This is a constant battle, but it helps me to keep learning and doing the best to be the master of my area," she says.

At work, her determination has helped her succeed. While working at Washington University School of Medicine in St. Louis, she led a project to address disparities in breast cancer diagnosis and treatment. The goal of the project was to understand social and environmental factors that cause African American women with suspicious mammograms not to seek treatment or to start or finish treatment to reduce breast cancer mortality. At the planning phase, one of her mentors told her the project would fail. She said no one would volunteer, but Faustine was determined to make it successful since it was her first lead project. The aim was to recruit 24 women. "A day before the workshop, I called participants using my personal phone and reminded them to call if they could not make it," she says. The project was so successful that even after leaving Washington University, Faustine still receives calls from community members wanting the university to consider working on other chronic diseases such as diabetes, hypertension, and so on.

Faustine, *right*, with her husband, Francis, and mother, Rose, receiving her US citizenship in 2016.

Inspiring others

When she taught during her PhD, Faustine says she would tell students it was okay to fail and not to be afraid to fail. "I think sometimes people give up way too early," she explains. "I see my intelligent failures as opportunities to discover new ways of doing things in addition to providing valuable new knowledge and experience for my growth as a scientist. There were instances that my very own research findings shocked me. The results weren't what we expected, and I wondered if I had wasted my time. However, those papers become my best findings." She adds, "We are

Fun fact!

Her favorite thing about GIS technology: "Making maps with fancy colors! I love colors, even though I am color-blind. I love GIS technology because a picture is always worth more than 10,000 words. Sometimes, I do not have to say anything— just show the results of maps and people get it. Very cool."

Faustine believes her greatest strength is her faith, determination, and a passionate belief that she can make it, and that nothing, not even failure, can stop her except herself.

human and live in an imperfect world. **Making meaningful mistakes is part of our growth process—hence, we should never be afraid to make intelligent mistakes."**

Because of her own experiences and passion for her work, she loves sharing her story with young people who think they cannot make it. "I love to encourage and motivate, especially those from disadvantaged backgrounds, that they can be whatever they desire with determination, perseverance, passion, and a goal in mind," Faustine says. She explains it's important for young women to know that life is not linear. **"A lot of people ask me how to get from start to end (success). I did not get here in a straight line. Things will not go according to your plan—but there are opportunities in the chaos."**

To encourage more young women to enter STEM fields and GIS, Faustine believes in a three-pronged approach. First, highlight the work and accomplishment of women in STEM, GIS, and research. **"It is very crucial to make ourselves visible or known (exposure is critical) so that young women are aware that STEM is not only for men or a particular racial/ethnic group,"** Faustine says. Secondly, she adds, "we must create the environment to mentor, nurture, inspire as well as invest in opportunities for young women (including women of color) to participate in STEM training in order to scale up apprenticeship." Lastly, she points to interventions at the institutional level. "Academia and other research institutions like the NIH should have a systematic approach/strategies/method to identify diverse young women and nurture them to become successful. **It is not enough to recruit. We must provide the environment for retention, growth."**

ABOUT THE ESRI PRESS TEAM

For writing and designing this book, special thanks to the following:

STACY KRIEG
Bringing it all together

Stacy Krieg is a senior acquisitions editor at Esri Press. Her experience working on volumes 1 and 2 of *Women and GIS* was so rewarding, and she feels privileged to get to know the women in this volume and honored to help them tell these inspiring stories. A native East Coaster, Stacy now lives in Redlands with her three children and dog and three cats.

CLAUDIA NABER
Learning about nature and women

Claudia Naber is an acquisitions editor at Esri Press. She truly enjoyed writing about such dedicated women and collaborating with the team at Esri Press. Claudia is also a caretaker and administrator at Animazonia Wildlife Foundation, a sanctuary for big cats rescued from threatening conditions in captivity or displaced in the wild.

ALYCIA TORNETTA
Searching for the heart of the story

Alycia Tornetta is an acquisitions editor at Esri Press. Coming from a fiction publishing background, she looks for the story in everything she does. She loved being able to tell the inspiring stories and achievements of the women in this book. She lives in Redlands

with her husband, two lazy cats, and two exuberant German shepherds.

JENEFER SHUTE
Finding the thread

Jenefer Shute is a developmental editor at Esri Press, currently based in the Hudson Valley of New York. Finding the common themes and threads in these women's stories has been an inspiration to her.

VICTORIA ROBERTS
Adding some color

Victoria Roberts is a graphic designer at Esri Press, where she lays out books, tinkers with data visualizations, and routinely wins trivia games. She loves the intersection of art and science and enjoyed the opportunity to showcase the stories of these exceptional women. In her spare time, Victoria travels, collects more books than she could ever possibly read, and makes illuminated manuscripts.

CAROLYN G. SCHATZ
Getting down to details

Carolyn Schatz is a senior editor and manager at Esri Press. A longtime journalist and newspaper editorial writer before her tenure here, Carolyn enjoyed getting to know these women through their stories and serving as editor of this book.

JENNA DEWITT
Editing with style

Jenna DeWitt is a copy editor for Esri Press and serves on the Esri editing team. She ensures Esri books, documentation, lessons, blogs, and more meet the highest standards and style guidelines. In her spare time, Jenna is a coleader of the lesbian, gay, bisexual, transgender, queer and questioning, intersex, aromantic, asexual, agender, and other sexual, romantic, and gender minorities (LGBTQIA+) employee resource group at Esri and supports equity and belonging efforts for all. She is thrilled to contribute her skills to a collection featuring such incredible women.

AND THE REST...

Storytelling map for this book created by product engineer Craig Carpenter. Marketing support provided by Beth Bauler, Sasha Gallardo, and Mike Livingston. With respect to the rest of the Esri Press team, including manager and publisher Catherine Ortiz, designer Monica McGregor, editors Mark Henry and David Oberman, and operations Sandi Newman and Jon Carter.